AUTOMOTIVE FUEL INJECTION SYSTEMS

A Technical Guide

**Includes information on
Bendix, Bosch, Cadillac, Chrysler, Ford, Kugelfischer,
Lucas, Marelli-Weber, Pierburg, Rochester, Zenith
and more . . .**

Jan P. Norbye

A FOULIS Motoring Book

First published 1988. Reprinted 1989, 1990 and 1992

This edition produced by arrangement with Motorbooks International Publishers & Wholesalers, Inc., Osceola, Wisconsin, USA.

Published by:
Haynes Publishing Group
Sparkford, Nr Yeovil, Somerset
BA22 7JJ, England

Motorbooks International is a certified trademark, registered with the United States Patent Office

The information in this book is true and complete to the best of our knowledge. All recommendations are made without any guarantee on the part of the author or publisher, who also disclaim any liability incurred in connection with the use of this data or specific details.

We recognize that some words, model names and designations, for example, mentioned herein are the property of various automotive and component manufacturers. We use them for identification purposes only. This is not an official publication.

British Library Cataloguing in Publication Data
Norbye, Jan P.
    Automotive fuel injection systems – new ed.
    1. Cars. Fuel injection systems.
    I. Title   II Series
    629. 2'53
    ISBN 0-85429-755-3

    ISBN 0-85429-761-8 (Motortrade edition)

Printed in England by: J.H. Haynes & Co. Ltd.

Contents

Preface

Since becoming a full-time journalist twenty-five years ago, I have reported extensively on the subject of fuel injection. The principle bears promise of technical superiority, and the early examples of fuel-injected cars were glamorous high-performance vehicles, as if to corroborate the impression of advanced technology.

My first experience with fuel injection occurred in 1954, when Rudolf Oeser, regional export manager of Daimler-Benz, took a 300 SL to Stockholm for a special exhibit. Sweden's economy was very strong, leading the company to conclude that there would be a Scandinavian market for that type of car. Support for this conclusion came from Sweden's lack of speed limits outside of built-up areas.

Before returning to Stuttgart, Rudolf Oeser took a side trip to Oslo, the capital of currency-restricted, speed-limit-ridden (43.5 mph [miles per hour] on the open highway!) Norway, where I was working on the monthly magazine of the leading automobile club.

We did not have good weather, and the demonstration was all the more convincing, as we rushed through a blanket of snow flurries whose accumulation on the cold roadway did not make the ride seem less hazardous, though it was conducted in perfect safety. My colleagues and I took turns riding in the passenger's seat with Oeser's expert demonstration driver at the wheel, a man who made the most out of every short bit of straight road we came upon, to the extent of twice being able to get into fourth gear, with joyful disregard for all speed limits. After looking at the engine, and seeing the ram pipes and the elegant fuel lines to the injectors, I came away thoroughly convinced of the merits of fuel injection.

Three months later, I was at Le Mans where I watched Froilan Gonzalez and Maurice Trintignant bring home their 4.9 liter Ferrari V-12 in first place. Their car was equipped with triple 46 DCF/3 Weber carburetors, and my feelings about fuel injection were brutally rocked back into proper perspective.

In the years that followed, I was able to see how the adoption of fuel injection for both racing cars and production cars was rejected or delayed by the excellence of Weber carburetors. Weber became in my eyes the worst enemy of fuel injection, surpassing the primitive applications of a superior principle in performance and efficiency by the sheer perfection of ancient technology!

Watching Juan Manuel Fangio win the French Grand Prix at Reims in July 1954, supported by Karl Kling and Hans Herrmann with their Mercedes-Benz W-196 Silver Arrows, did provide me with a boost for the belief that fuel injection was bound to triumph in the long run.

In December 1961, I had a chance to ride in the 300 SLR (powered by a three-liter version of the M-196 fuel-injected straight eight) as Rudolf Uhlenhaut, chief engineer of car development, toured around the banked tri-oval Daytona Speedway at speeds up to 145 mph.

By mid-1963, I had racked up impressive mileages with the fuel-injected Peugeot 404 convertible, the Mercedes-Benz 220 SE, the new 230 SL and the Maserati 3500 Iniezione, each one stating a convincing case for fuel injection, each in its own way. But a few years later, Weber snatched back its previous monopoly on fuel mixture preparation hardware for Maserati engines, much to Lucas' dismay.

The switch to fuel injection was not a one-way street, as many had expected. Aston Martin and Triumph were other makes that reverted to carburetors after brief periods of producing fuel-injected models. These happenings were highly influential in the maturing of my own philosophy about fuel injection. Much as my natural enthusiasm leaned in favor of fuel injection, it was tempered by my admiration for a new and better carburetor.

The advent of electronic fuel injection in 1965 stirred things up in my mind, and new experiences soon came my way, including 6,000 miles in three weeks with a Citroen DS-23 Injection, a streamlined four-door sedan that would cruise at 115 mph for hours on end, with nothing more than a 140 cubic inch four-cylinder engine under the hood!

There were also the Mercedes-Benz 350 SL and 450 SEL, Alfetta from Alfa Romeo, Porsche 911, Peugeot 504 convertible, Saab 99 GLE and a host of others, either furthering the electronic revolution or exploring simpler, more conventional means in the form of Bosch's K-Jetronic.

For years, the trend was clear and strong, but after 1975, some flexing has set in, owing to the need for lower-cost systems. This quest has produced impressive new solutions in the form of single-point injection and computer-controlled carburetors.

Where do we go from here? Electronics are taking over an increasing number of duties, but let's not get tricked into thinking that controlling the brakes, differential, transmission, low/high beam light switch or sun visor angle is going to do much to advance the efficiency of the combustion process.

There is plenty of room, however, for new ideas in fuel mixture preparation. The potential of pneumatic injection has not been fully explored, and a good case can be made for incorporating water injection as an integral part of the fuel injection system. We must also face the possibility that an advanced fuel injection system might one day bring the two-stroke engine back into the realm of automotive realities. Finally, this book stops short of discussing alternative fuels, but there is much going on around the world that is likely to offer opportunities to inventors of fuel systems in the future. Happy brainwaves!

Jan P. Norbye

Introduction

Priorities in automotive engine design have undergone two major changes in the past twenty years. The emphasis used to be on maximum horsepower, consistent with the basic requirements of reliability and long life. Then came the Clean Air Acts of 1966 and 1970, which made an absolute priority of antipollution measures. After 1973 and the rapid quintupling of crude oil prices, priorities were redirected toward fuel conservation, which puts a premium on thermal and mechanical efficiency in an engine.

Horsepower, cleaner exhaust, lower fuel consumption—that just about sums up what we are looking for in our next car. And we can get it, we are told, from one single thing: fuel injection.

Many engineers now hold the opinion that fuel injection will be a vital means of making further progress in operating efficiency without abandoning the gains that have been made in the emission control area. Others refuse to give up the carburetor, and hint that fuel injection is merely a passing phenomenon, a fad.

It seems disrespectful to talk of fads in automotive engineering, but it is true that the industry is not immune to the vagaries of fashion—examples abound. Now it's front-wheel drive. Suddenly, everybody in Detroit wants to make front-wheel-drive cars, after stubbornly resisting the European trend for generations. Before that, it was disc brakes and radial tires. Next, it may be fuel injection.

Call it a wave, to get away from the insulting term *fad*. The mass movement toward technical innovation is partly explained by cost considerations. The desire for engineering progress has always been tempered, in mass production, by the specter of higher costs.

Of course, all that's new isn't necessarily better. The way for the broad adoption of disc brakes was paved by the pressing need for better brakes. Making substantial improvements in drum brake systems would cost a lot of money and alter the balance in the comparison with disc brakes, in favor of the latter. And as engineers know, after some experience with an initially expensive device they invariably find ways to make substantial cost cuts, either by simplifying the product or by developing new production methods, or by a combination of both.

Rarely does a new feature show a total and indisputable superiority. It is possible to make very good drum brakes or very good bias-ply tires, for instance.

What happens is something like this: Coupled with the change in the industrial supply situation that occurs when a major car manufacturer undertakes a wide-scale change in the product, the publicity and marketing arms of the producers go into action to persuade the public that the change is, in fact, progress.

It's easy to sell a superior principle, such as disc brakes or radial tires, and customers readily pay the higher price, in return for the advantages obtained. Fuel injection has often been hailed as such a superior principle, and abundant evidence exists that car buyers were and are willing to accept the higher price. But when you closely examine the functions of a fuel injection system, and those of a carburetor, this superiority is not so clear-cut. It could be questioned, or even challenged.

The differences lack contrast. The carburetor mixes fuel with air to form a combustible mixture. So does a fuel injection system.

Are there really different principles at work? If you go deep enough into the details, yes. In the carburetor, fuel is sucked into the air, while fuel injection squirts it into the air. It comes down to a choice between suction and pressure-spraying. The distinction is fine. If you have a system that works with very low pressure injection, who can say that suction forces play no part in getting the fuel out of the nozzles?

But what about all those other claims enlisted in support of proving the superiority of fuel injection? It has highly precise fuel metering, you have been told. That depends entirely on the type of measurement control that's built into the system. Accurate timing of fuel delivery? Not all systems have timed injection; some spray continuously.

Modern carburetors have been brought to a high state of perfection, and there is no reason to expect that they have reached their final stage.

Have the engineers let themselves be hypnotized by the vague promises of fuel injection? No, that's not the answer. Disagreement among engineers as to the merits of fuel injection probably stems from differences in their perceptions of the qualities and short-comings of the carburetor.

So many compromises are made in the carburetor that none of its functions can be performed optimally. Thus, fuel injection may look attractive immediately on the strength of the realization that doing away with the carburetor eliminates its compromises and opens the way for more efficient engine operation. But getting the full benefits of fuel injection depends on designing and developing the engine to take full advantage of its promises.

Take the question of power gains, for instance. Why does fuel injection enable us to take more power out of an engine without increasing fuel consumption? Because fuel-injected engines will run smoothly with far higher compression than can most carburetor engines.

If automobile engines ran at a constant speed, the carburetor would not have a difficult task. As you know, however, they have a wide speed range, and they must operate under variable load.

According to Harry Mundy, once Jaguar's engine designer, "The basic disadvantage of the carburetor is that there must be a constriction at the venturi, whose function it is to increase the velocity of the incoming air, thereby creating a vacuum to draw the fuel through the jets from the float chamber. Such a restriction limits the amount of mixture passing into the cylinders, and results in the power falling off at higher engine speeds. This can be overcome by fitting larger carburetors, but then the difficulties are transferred to the lower end of the speed scale, and flexibility is lost."

Intake manifolds are not, and perhaps cannot be, designed to facilitate things for the carburetor. The runners twist and turn, for their shape is determined mainly by the space available for them.

After leaving the carburetor, the mixture has to travel a sinuous route, with changes in flow velocity and pattern that are different for each port. Because of these differences, it is not possible to obtain uniform mixture distribution, and some cylinders run with a richer mixture than others.

That does not give the complete picture, however. The carburetor also must be made to overcome three other fundamental problems of fuel mixture preparation and engine operating conditions.

One difficulty is cold-starting. Another is the need for transient enrichment during acceleration. The third would not exist if cars traveled only in a straight line. However, roads have curves and streets have corners, and any change in the car's direction sets up a centrifugal force that affects all parts of the car and everything carried in it, including the fuel in the float bowl.

Fast cornering tends to force the fuel in the float bowl to climb up the wall, thereby raising the float to block further delivery. Result: fuel starvation. The driver experiences a sudden loss of power just when it is most wanted.

The third problem can be prevented by careful attention to float action at the design stage, which does not necessarily add to the cost of the carburetor. But the first and second difficulties require additional mechanisms for solution.

Carburetors must be designed to provide a richer mixture for starting. With cold air in a cold manifold, atomization is poor. A portion of the fuel separates from the air and clings to the walls of the manifold. The carburetor must make sure that enough fuel goes into the engine to have a mixture that can be ignited in the cylinders.

To do this, it has a choke mechanism, which can be operated manually or automatically. But in neither case can it be assumed that it responds with any accuracy to the operating conditions.

Then there's the problem of speeding up. Instantaneous opening of the throttle valve tends to lean out the mixture because the speed of the fuel flow doesn't increase as quickly as air velocity. To avoid bucking or snatching, stumbling or hesitating, during sudden acceleration, an accelerator pump feeds additional fuel into the venturi. The accelerator pump, built into the carburetor, sprays a jet of fuel into the onrushing air at the critical moment. This pump is usually a simple plunger pump connected to the throttle linkage. The fuel from it is metered into the airstream in a steady discharge of short duration.

With fuel injection, the need for an acceleration pump, a choke mechanism and a special means to ensure continued fuel flow under high-lateral-force cornering conditions can be included as integral parts of the system and made to function automatically.

We must guard against thinking of fuel injection in terms of a rigid system and specific hardware. Instead, we must look at fuel injection as an operating principle that can be applied to several types of different systems. New variations are continually appearing, and the ultimate form of fuel injection may not yet exist.

Even for the time being, the fuel injection engineer has many difficult choices to make. Where to inject? and When to inject? are the two primordial questions to be answered. One could inject (1) inside each cylinder, (2) into each port or (3) farther upstream in the intake manifold.

The location of the injector nozzle dictates the timing to some extent. With direct injection (inside each cylinder), injection must occur in a short period during the compression stroke.

With port injection, timing is not critical. If it is timed, it should coincide with the valve opening period. Upstream injection is not sensitive to timing, but involves shortcomings in the control of mixture strength to the individual cylinders.

With timed port injection, the duration is partly a function of the quantity to be injected. With high-speed engines, a physical problem arises in that extremely minute quantities of fuel must be injected (and given time for atomization) in periods that can be measured only in milliseconds.

Direct injection requires a high-pressure pump which, aside from being noisy, consumes energy to pressurize the fuel. For production cars, the engineers' attention has been focused on other systems, working with lower-pressure pumps consuming less energy. The lowest fuel delivery pressures are obtained with electronic fuel injection systems. In some companies, borderline systems that borrow from both carburetors and fuel injection, controlled by electronic means, are under investigation. In these companies, a new term has come into usage: electronic fuel management.

That may be the wave of the future. The following text examines the principles of mixture preparation, the combustion process, and past and present fuel injection systems, so as to give a solid background of the whole subject, from the basics to the frontiers of technology.

1

Injection or carburetion?

As demonstrated in the introduction, the main attractions of fuel injection are to be found in the shortcomings of the carburetor. Thus, the carburetor is a logical starting point for any comprehensive examination of fuel injection principles and hardware, and a detailed look at its construction and various functions is necessary.

Since the carburetor is, in a sense, the guard at the gate, letting air in or preventing it from entering, it's easy to see that the carburetor is the first link in controlling how the engine breathes.

In theory, during the intake stroke the piston should draw into the cylinder a volume of mixture at atmospheric pressure equal to the cylinder displacement. In practice, the quantity of gas actually drawn in is always less than the theoretical amount.

The ratio between the theoretical amount and the actual amount is called *volumetric efficiency*. Its usual value for a modern engine is eighty to eighty-five percent at full throttle. Poor volumetric efficiency does not necessarily hurt fuel economy but results in low power output relative to engine size. The principal reasons for failing to reach 100 percent are as follows:

1. Restrictions in the carburetor, and bends in the intake manifold and porting system, limiting gas flow to the cylinders

2. Heating of the incoming charge by a hot intake port, or by other hot parts in proximity to the intake manifold, causing the air-fuel mixture to expand before entering the cylinders

3. Hot exhaust gases remaining trapped inside the cylinder after the exhaust stroke

The basic idea of the carburetor is to have an air passage with automatic fuel feed that is self-regulated to suit the mass airflow as measured through a venturi. In principle, a carburetor consists of a fuel reservoir or float chamber and a fuel jet leading to the main

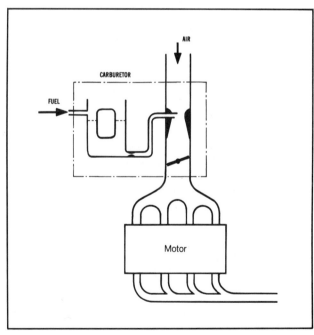

Carburetor principles: Nothing is measured, but fuel metering maintains a nearly constant proportion with airflow because of its dependence on manifold vacuum. The sketch also makes the point that mixture for all cylinders is prepared together, upstream of the intake manifold.

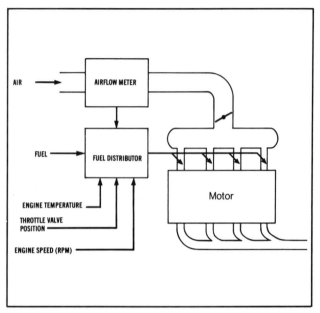

Fuel injection principles: The basic difference with a popular, modern Bosch injection system from the typical carburetor is that air and fuel are measured separately. Fuel metering takes account of mass airflow, engine speed, throttle position and engine temperature as the essential control parameters. In some systems, further parameters are added. Fuel is delivered individually to each cylinder, in uniform quantities for any given set of conditions. That saves fuel and gives more power.

venturi, or air passage, equipped with a throttle valve. A constant level is maintained in the reservoir by a float that opens or closes the fuel valve through a simple linkage. Fuel supply is provided by the action of the fuel pump, bringing raw gasoline from the tank.

The carburetor works by mixing the charge of fuel and air and distributing this mixture to the cylinders through the intake manifold. The mixture must be rich enough to ensure that the cylinders located farthest away from the carburetor get enough fuel. The others, consequently, tend to run overrich. That's a big handicap for both fuel economy and emission control.

All carburetors have a throttle valve to control the volume of intake air. The mass of air that goes into a carburetor regulates engine speed and can therefore be used to regulate engine power. The throttle valve is linked to the accelerator, or gas pedal. The throttle valve is a butterfly valve that consists of a disc mounted on a spindle. The disc is roughly circular, and it has the same diameter as the main air passage in the carburetor. It is located at the bottom of the carburetor, between the jet nozzle and the intake manifold. The throttle spindle is connected to the accelerator in such a manner that when the pedal is depressed, the valve opens. When the pedal is released, the valve closes. The fuel jet is fed from the reservoir and pro-

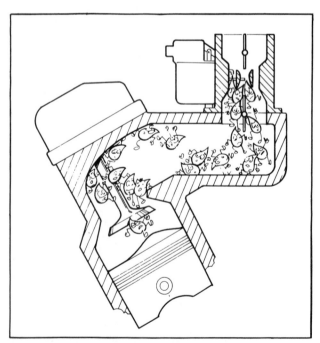

With a carburetor, fuel droplets are sucked in by manifold vacuum for mixing with the incoming air upstream of the manifold flange. There is a big risk of wetting the manifold walls with the fuel droplets, which upsets the balance of the mixture.

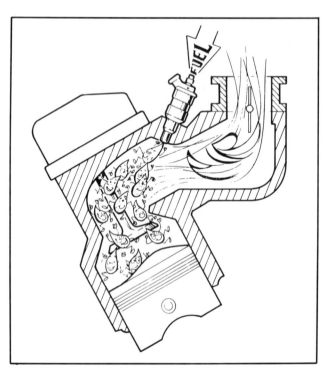

With fuel injection, the fuel droplets are squirted into the inrushing airstream under pressure. With the injector mounted close to the intake valve, there is little or no chance of wetting the manifold walls. All the fuel goes into the cylinder without delay, allowing very precise control over the air-fuel ratio.

jects into the narrowest part of the venturi where air velocity is highest. Gasoline does not normally flow out of the jet nozzle but must be trickled out by the airflow.

This trickling supply of fuel is assured because there is higher pressure in the carburetor bowl than in the engine manifold. Of course, it's not practical for the carburetor bowl to be pressurized. It is left at atmospheric pressure since it is much simpler to establish a pressure difference by creating a pressure drop or partial vacuum on the engine side of the carburetor.

But the fact remains that fuel delivery in a carburetor tends to lag behind the throttle motion, mainly because of surface tension and inertia in the fuel.

The basic carburetor operates when the throttle valve is fully open or partially open, but not when it's closed. But no driver wants the engine to stop every time his or her foot leaves the accelerator. A car with such an engine would be very tiring to drive under the easiest conditions, and almost impossible to drive in heavy traffic, with frequent stops and starts. To keep the engine running even when no power is needed, the idle circuit was added inside the carburetor. The idle jet admits fuel on the engine side of the throttle valve. Additional air is mixed with this fuel through an air bleed. The result is an entirely separate carburetor circuit which operates only when the throttle valve is closed.

Airflow through the carburetor is assured by the pumping action of the pistons. The downward movement of the piston in the intake stroke creates a partial vacuum in the cylinder. Air-fuel mixture in the intake manifold rushes in to fill the vacuum, and the gas flow set up by the pressure drop draws fresh air into the carburetor.

A pressure drop is instigated by a device called a venturi. The venturi is a constriction in the carburetor throat. The only way a constant amount of gas can flow through a tube that gets narrower is by speeding up. And according to Bernoulli's law, an increase of air velocity through the venturi will be accompanied by a reduction in pressure. The venturi is placed so that this pressure drop will be highest in the area near the jet nozzle. Fuel will flow from the float chamber to the jet orifice because the pressure on the surface of the fuel in the float chamber is atmospheric, while the jet

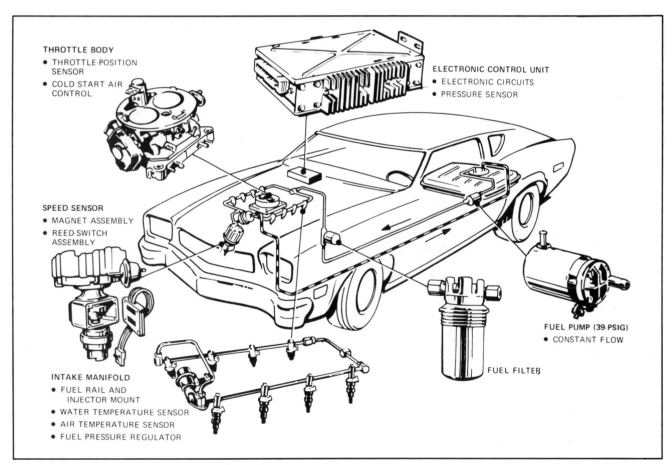

Elements of an early-model electronic fuel injection system. This system provides timed injection of accurately metered amounts of fuel into the port areas.

outlet is in an area of partial vacuum. The fuel flows out of the jet as fine spray and is carried along by the entering air.

To obtain proportional and automatic fuel feed into the airstream, the carburetor has to maintain the correct air velocity past the fuel jet nozzle under light as well as heavy loads.

The venturi's efficiency depends on its length-to-diameter ratio. The amount of fuel drawn into the engine depends on how big the pressure drop is. The smaller the fuel particles as they leave the carburetor, the more easily and thoroughly they will mix with the air and vaporize in their passage through the intake manifold to the cylinder. The dimensions of the air supply pipe and the jet are carefully proportioned to give the correct air-fuel ratio. The problem is that this ratio is not constant, and the carburetor cannot accurately adjust it to changing needs.

In any case, for proper ignition and waste-free and complete combustion, the mixture should be a homogeneous vapor containing no liquid fuel. That is conditional on thorough mixing—and it is a factor that also must be taken into account for fuel injection systems.

With carburetors, the thoroughness of the mixing process is mainly determined by the distance from the carburetor flange to the inlet valve port, and the gas flow velocity. The rate of fuel flow at the nozzle increases faster than the pressure drop in the venturi. This means that the engine would run richer and richer as speeds rise, unless the carburetor could correct the air-fuel ratio. Air correction is achieved by introducing air into the fuel supply before it leaves the nozzle. This is usually done with an air bleed.

The most common type of air bleed is an emulsion tube, which is a short tube with holes drilled across it. It is located in a fuel well inside the carburetor. As the pressure drop increases, fuel flows faster. This lowers the level in the fuel well and uncovers more holes in the emulsion tube. As a result, additional air is bled into the mixture, and the formation of an overrich mixture is prevented.

The degree of atomization varies greatly with changes in engine speed and load (as measured from the throttle opening). A large-diameter venturi would be best for full-power operation. A small-diameter venturi would be best for part-throttle operation. A small-diameter venturi also offers fuel economy advantages, but while it tends to provide adequate acceleration potential, it inevitably entails a slight loss in top speed.

Many manufacturers increased venturi size to provide the full airflow needed by their engines for maximum power output. Then it was found that there was a practical limit to maximum venturi size for acceptable low-speed vehicle operation. This limitation led the industry to the modern two- and four-barrel carburetors.

Barrel is a popular term for carburetor throat. There is one venturi in each throat. A two-barrel carburetor has a primary venturi for part-load running and a secondary venturi for full-throttle running. The four-barrel carburetor has two primary and two secondary venturis. Only the primary venturis

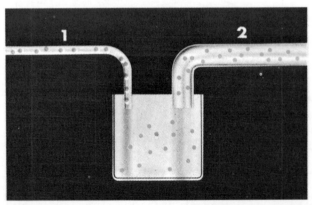

Air-fuel ratio: The combustion process needs oxygen from the air to burn the fuel. The theoretically ideal proportion for complete combustion of one part of gasoline is 14.7 parts of air. That is called the stoichiometric ratio. If the fuel component is higher, the fuel is not fully utilized, and the amount of unburned hydrocarbons in the exhaust system is increased. With less fuel, compared with the amount of air, the power drops and the combustion process slows down, which will cause engine temperature to rise.

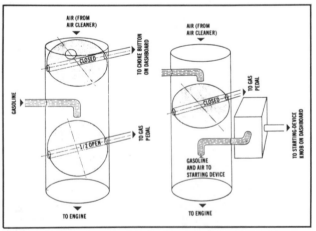

The cold-start problem: Whether the engine is equipped with a carburetor or fuel injection, it demands a richer mixture for cold-starting. The problem is worse in carburetors, since the cold air prevents fuel atomization and a portion of the raw fuel remains in the manifold. The sketch shows two types of aids to the carburetor: left, the manual choke (which can be made automatic); and right, the special starting device of a Solex carburetor. The choke blocks off air, while the starting device adds a richer mixture downstream from the throttle. Starting enrichment in fuel injection systems is provided partly by the basic delivery, and partly by an upstream starting valve.

are used under part-load conditions. The throttle valves on the primary throats are linked to the valves in the secondary throats. When the primaries are, for instance, half-open, the secondaries only begin to open. At full throttle, all throats are fully open. This assures a certain economy under light-load operation as well as maximum gas flow and correct air-fuel mixture for full-power operation.

Theoretically, the ideal air-fuel ratio is 14.7 parts of air to one part of fuel. However, experience has shown that this is true only in steady-state operation, such as turnpike cruising, with minute variations in throttle valve angle and engine revolutions per minute.

For part-throttle, light-load operation, a lean ratio of 16.0 to 16.5:1 will be adequate. This ratio must be enriched for speeding up, approaching 12:1 for full-throttle acceleration. A ratio of 10 or 11:1 will be best for a hot engine at idle. But for cold-starting, the air-fuel ratio must be as rich as 3 or 4:1.

Air and fuel don't mix very well in cold weather. Only the lightest portions of the gasoline help form a combustible mixture. For that reason, the mixture must be richer when the engine is cold and the ambient temperature is low.

Mixture enrichment is achieved by a choke mechanism. The choke is a special valve placed at the mouth of the carburetor so that it partially blocks off the entering air. By severely reducing air supply, vacuum is drastically increased at the venturi, causing the fuel flow to speed up. This results in a richer mixture. The engine starts—but will it keep running? Blocking off part of the air intake tends to stop the engine, doesn't it? Yes. When the engine is cold and operating with the choke valve shut, the total volume of entering mixture is so small that the engine has a tendency to stall. Something had to be invented to keep it running.

This much-needed invention is a simple thing called the fast-idle cam. This is a part of the choke mechanism, hooked up to the throttle linkage to give a fast idle as long as the choke is in function.

The *manual* choke is a knob on the dash, usually a push-pull. The main thing to know about it is to remember to push it back in when the engine has reached operating temperature.

Automatic chokes rely on engine heat for their operation. The choke valve is operated by a thermostat controlled by exhaust heat. With a cold engine, the choke valve will be closed for starting. As the engine warms up, the exhaust heat will gradually open the choke valve.

The thermostat consists of a heat-sensitive spring. The most common type is a bimetallic coil spring, much like a watch spring in appearance. This spring is fixed at its center and attached to the choke valve by a linkage at its circumferential end. The spring is contained in a housing, and a tube connects the housing to the exhaust manifold. When the engine is cold, the spring contracts to close the choke valve. As soon as

the engine begins to warm up, heat from the manifold makes the spring unwind itself, and the choke valve gradually opens.

The intake manifold incorporates a special passage for exhaust gas to warm the incoming air-fuel mixture and improve atomization on cold starts. After the engine is warm, this heating is undesirable; it would cause the fresh mixture to expand before entering the cylinder, which would reduce the volumetric efficiency of the engine. Therefore, a heat-riser valve is provided to direct exhaust gas flow according to engine temperature.

The heat-riser valve is located on the exhaust manifold. It regulates the amount of exhaust gas allowed to pass through the intake manifold. A bimetal spring attached to the control valve shaft gradually restricts the amount of exhaust gas flow to the intake manifold by slowly closing the valve. When proper operating temperature is reached, the exhaust gas is routed to the muffler and out the tailpipe.

Before the advent of emission controls, fuel delivery was considered accurate enough if variations in the air-fuel ratio across the full range of airflow mass and velocity were kept within plus or minus five percent of the nominal setting. By 1969, nothing above a three-percent variation could be tolerated, and by 1972, the margin was reduced to between 1.0

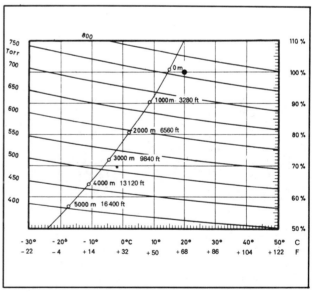

Engine performance drops at high altitude, when the air gets "thinner." A reduction in the weight of the air sucked into the engine leads to less oxygen being available for the combustion. That's a natural phenomenon, whose effects can be plotted by the aid of mathematical formulas once the temperature and barometric pressure are known. The altitude curve here (starting at the top) makes it possible to read off the average loss in performance even if these values are not known. At an altitude of 2,000 meters above sea level, for instance, engine performance is only about 81 percent of full power.

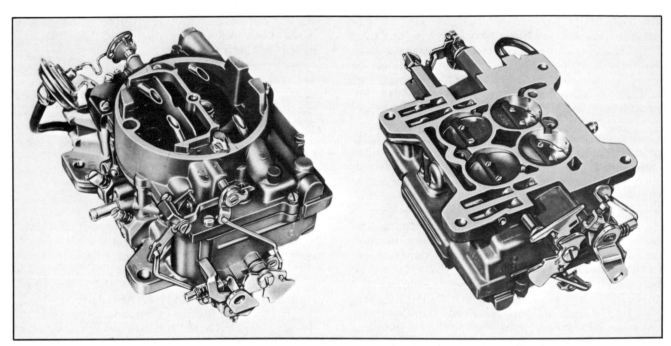

The four-barrel carburetor is a maze of mechanical, pneumatic and hydraulic (fluid-flow) systems interacting to satisfy the engine as well as the carburetor principle will allow. Some fuel injection systems may be less expensive.

RUNNING ADJUSTMENTS

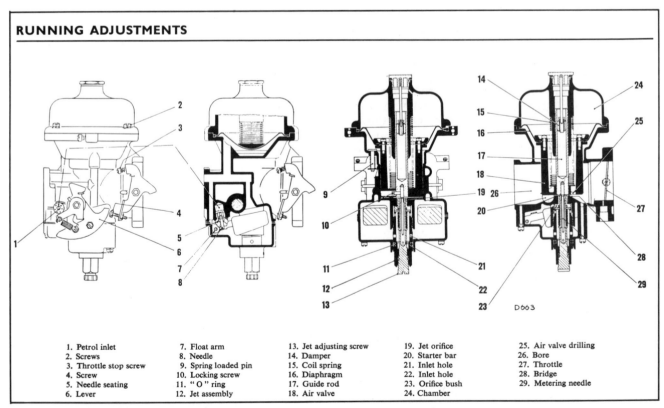

D663

1. Petrol inlet	7. Float arm	13. Jet adjusting screw	19. Jet orifice	25. Air valve drilling
2. Screws	8. Needle	14. Damper	20. Starter bar	26. Bore
3. Throttle stop screw	9. Spring loaded pin	15. Coil spring	21. Inlet hole	27. Throttle
4. Screw	10. Locking screw	16. Diaphragm	22. Inlet hole	28. Bridge
5. Needle seating	11. "O" ring	17. Guide rod	23. Orifice bush	29. Metering needle
6. Lever	12. Jet assembly	18. Air valve	24. Chamber	

The air valve or constant-vacuum carburetor: The whole air valve assembly rises or falls according to the strength of the vacuum in the depression chamber (24). The piston (18) controls the cross-sectional air inlet area, its motions being damped to avoid overreactions to rapid flooring of the accelerator pedal. An air valve carburetor has only one air passage; to give proper mixture distribution, a multi-carburetor setup may be required for multicylinder engines.

and 1.5 percent. Such precision is extremely difficult to obtain in mass-produced carburetors.

The complications of making the carburetor fit the requirements of the engine do not end there. All cars with automatic transmission have a dashpot to keep the engine from stalling when the accelerator is suddenly released. The dashpot is actuated by an arm of the throttle lever when the throttle is closed. It cushions the closing of the throttle, letting fuel metering abate gradually instead of being shut off all at once.

Automatic altitude compensation would be quite expensive to include in a carburetor. Because of rarefied air at high altitude, a normal setting would give an overrich mixture. Statistically, few cars move back and forth between the two different environments of low and high altitude, and the industry has been content to deliver cars with different carburetor settings according to destination.

The carburetor, despite all its complications—air bleeds, correction jets, acceleration pumps, emulsion tubes and choke mechanisms—remains a compromise. The additional costs of better carburetor design compromises are helping to push the industry toward fuel injection.

But as shown later, fuel injection systems are not, themselves, free of compromise. Two arguments remain, standing unchallenged, in favor of fuel injection. One is the engine's ability to run with higher compression ratios when the fuel is injected.

The other argument—and some experts say it is the only real advantage of fuel injection—lies in the freedom of intake manifold design. With carburetors, manifolds must be designed to encourage fuel atomization and discourage raw fuel droplet accumulations on the runner surfaces. To this end, manifold heating must be provided.

For engines with carburetors, the intake manifold has two main duties:

1. It must deliver equal quantities of fuel and air mixture to all cylinders.

2. It must ensure that the mixture possesses the same chemical and physical characteristics in all cylinders.

Gases are both elastic and viscous. With high gas velocities through imperfect passages, the gas composition will break up and gas flow will change from smooth to turbulent. Only individual particles will maintain theoretical velocity.

When the cross-sectional area of the manifold is too large, gas velocity drops below a critical point where drops of raw fuel begin to settle on the manifold walls. This changes the proportion of air to fuel as metered by the carburetor.

The manifold cross section is usually as large as possible for ideal power conditions, without reaching the critical point. Inner walls of the manifold are made as smooth as possible because gas flow separation can be caused by roughness in the surface. On the other hand, this tends to promote fuel droplet deposition. Sharp corners in a manifold are sometimes deliberately built in. They can cause a film of fuel to tear loose from the wall and reenter the airstream, but since this occurs at random, it can also cause a momentary upset of the air/fuel ratio.

Depending on the exact location of the injector nozzles, the mixture control considerations do not apply to manifolds for fuel-injected engines. With fuel injection, the engineer has a free hand to exploit ram effects to the maximum.

Ram is an air momentum effect. Once gas flow has been started and accelerated, there will be inertia in the gas when its flow is suddenly stopped by a closed valve. This inertia can be utilized to set up pulses in the manifold runners that will force more gas past the intake valve than would be possible by suction only. Intake pipe length governs ram effect. There is an ideal length for each cylinder for each engine speed, and all manifolds are a compromise between optimum high-speed and best low-range gas flow characteristics.

2

Fuel properties

When this text discusses fuel injection in comparison with carburetors, the fuel used is the same: gasoline. Fuel injection is also used on all diesel engines, but diesel engines operate by different principles, and their injection equipment and fuel properties are not discussed here.

Fuel injection is fully compatible with gasohol and certain other alternative or future fuels, but existing systems are not designed for those. The following discussion is restricted to gasoline as you buy it at the pump.

The first fact that needs to be faced about this fuel is that raw gasoline will not burn. It must first be changed from liquid to vapor and supplied with an adequate volume of air to ensure that enough oxygen to support combustion enters the engine. Gasoline is a hydrocarbon, made up of about fifteen percent hydrogen and eighty-five percent carbon. There are other ingredients that are discussed in more detail later, but this is all you have to know about gasoline to understand its use as motor fuel. Air is a mixture of twenty-one percent oxygen, seventy-eight percent nitrogen and one percent other gases. But only the oxygen combines with the gasoline.

Before gasoline can be thoroughly mixed with air in proper proportions, it must be broken up, or atomized, into finely divided particles. But the engine won't run satisfactorily if it only gets an air-fuel mixture of the same proportions all the time. The mixture must be adjusted to speed, load conditions and temperature. And the total volume as well as the air-fuel ratio must be continuously adjusted to these conditions.

Gasoline is obtained by refining crude oil. Crude oil as it comes from the ground is a mixture of thousands of different chemicals, which range from extremely light gases to semisolid carbon-containing materials such as asphalt or paraffin wax. The gases are dissolved in the other components of the crude oil because of the extreme pressure at which petroleum is stored in the ground.

Water weighs 8.33 pounds per gallon. Crude oils vary between 6.5 and 8.3 pounds per gallon. As a liquid, crude oil may be as thick and black as melted tar or as thin and colorless as water. Its characteristics depend on the particular oil field from which it comes. Crude oils contain sulfur compounds in varying amounts. Sulfur is undesirable in motor fuel because it gives rise to bad odors, is corrosive to engines, is poisonous to some catalysts and may reduce the efficiency of antiknock compounds. A maximum amount of sulfur is removed as part of refining.

The refining process begins with distillation of the crude oil. That is followed by vacuum fractioning, which separates light and heavy oils, gas oil and bitumen. Refining also includes such supplementary operations as the stabilization of the gasoline to eliminate condensible elements, refining of white spirits, extraction of oils by means of solvents, and processing of paraffin and bitumen.

The cracking process is also important. Higher-boiling hydrocarbon molecules can be broken down, or *cracked*, into lower-boiling ones by subjecting them to extreme temperatures. Thermal cracking was the natural and simple way to do this. However, the original method has been almost completely replaced by catalytic cracking, which gives greater freedom of production for a refinery.

Cracking is fractional distillation combined with other operations. After an initial stage of physical purification, the oil is subjected to fractional distillation at atmospheric pressure in columns some thirty meters high. These columns are heated at the base, and temperatures range between 350°C and 70°C. The most volatile products run to the top. From top to bottom are gasoline, kerosene, gas oil, fuels and lubricants.

Catalytic cracking is a reforming process which transforms heavy essences of first distillation into light ones with a higher octane number. In catalytic reforming, there are molecular rearrangements as well as molecular splitting. The performance of gasoline is determined mainly by its volatility (tendency to boil and vaporize), by its antiknock quality, and by its cleanliness and stability. These characteristics are determined by the refiner's selection and processing, plus the additives and inhibitors the refiner adds.

Gasoline must form vapors at a low temperature to assure easy starting. It must vaporize at an increasing rate as carburetor and manifold temperatures rise to assure fast warm-up, smooth acceleration and even fuel distribution among the cylinders. The vaporizing characteristics must be in keeping with the climate and altitude where the gasoline is to be used, to prevent vapor lock and fuel boiling inside carburetors, pumps and fuel lines.

Gasoline should contain few extremely high boiling hydrocarbons to ensure good fuel distribution and freedom from crankcase deposits and dilutions. A high antiknock quality (octane number) throughout its boiling range is needed to assure freedom from knock at all engine speeds and loads. Gum content must be low to prevent valve sticking, carburetor deposit difficulties plus deposits inside engine and intake manifolds. Gasoline must also have good stability against oxidation to prevent deterioration and gum formation in storage.

The antiknock property of a gasoline is indicated by its octane number. The octane scale was created by giving the number zero to heptane (C_7H_{16}) and the number 100 to isooctane (C_8H_{18}). Numbers between zero and 100 indicate the proportion of each if the two are mixed. The octane number of a gasoline is determined by a test comparing it with a mixture of heptane and isooctane. If the gasoline shows the same tendency to knock as a mixture containing six percent heptane and ninety-four percent isooctane, its octane number is ninety-four.

There are two ways of establishing a fuel's octane number. In the *Research Method*, the test engine is run under closely controlled conditions of speed, air intake temperature and ignition timing. In the *Motor Method*, the test engine is run with variations in speed, air intake temperature and ignition timing. The difference between Motor and Research Method ratings for the same fuel is called *sensitivity*.

It is impossible to obtain more than 100 percent isooctane in a reference fuel blend, but some fuels have ratings above 100 octane. Then the reference fuel becomes isooctane plus a certain amount of tetraethyl lead. The knock value of such fuels can be defined as milliliters of tetraethyl lead per gallon of isooctane.

The fuel injection specialist must take close account of an engine's octane requirement. Octane requirement can be defined as the octane number that produces a given level of knock resistance during acceleration with a throttle opening that is known to produce maximum knock.

Octane requirement depends on many variables. Identical vehicles coming off the assembly line can have differences of up to ten octane numbers. The conditions that dictate octane requirement are atmospheric pressure, relative humidity, air temperature, fuel characteristics, air-fuel ratio and variations in air-fuel ratio between individual cylinders in the same engine, oil characteristics, spark timing, distributor advance curve, variations in timing from one cylinder to another, intake manifold temperature, water jacket temperature, condition of coolant antifreeze, type of transmission and the presence of hot spots in the combustion chambers.

Antiknock compounds are usually lead-based. Since lead is a poison, the Environmental Protection Agency has imposed strict limits on the lead content of gasoline. Some European countries have no lead restrictions. Sweden and West Germany have limits for lead content, but they are less severe than those of the United States and Japan.

Where leaded fuels are still produced, scavengers are added to remove the small amount of lead compounds that might remain in the engine after combustion. These scavengers include bromine and chlorine. They convert the lead compounds to lead bromine and lead chloride. These are gaseous at the temperatures prevailing inside the engine. The scavengers remove practically all lead and dispose of it in the exhaust fumes. When lead is found in the combustion chamber deposits, it is mainly because organic residues from burned oil or fuel have acted as blinders for them.

Antioxidant chemicals are added to gasoline to protect against formation of gum and peroxides. Numerous chemicals have been used as antioxidants. Until recently, phenylene diamine, aminophenols and dibutyl-cresol were those most widely used. Today the most used and most effective antioxidant is a new family of ortho-alkylated phenols.

Metal deactivators, usually an amine, are often used along with antioxidants. They prevent trace amounts of copper (picked up from the piping or the engine fuel system) from acting as a catalyst for the formation of undesirable materials in the gasoline. Other additives include anti-icers (alcohol) to prevent carburetor icing and fuel line freeze-up, and antirust agents. There are detergents to keep carburetor parts clean—and phosphorous additives to combat surface ignition and spark plug fouling. Dyes are added to identify leaded gasoline.

Gasoline is carried in the fuel tank, which is usually positioned in the rear part of the car, under or ahead of the trunk floor. Fuel level in the tank is sensed by a float, coupled to a sender unit, connected to a gauge on the instrument panel.

Filtration of the fuel is necessary to keep water and dirt from entering the engine. Water is the main problem; moisture in the air inside the tank will condense and sink to the bottom of the tank. Some cars have a filter inside the tank, at the main fuel line where it leaves the tank. Some cars even have coarse filters in the filler neck.

The main fuel filter is usually integral with the fuel pump. Fuel passes through a filter screen before passing through the outlet. Dirt and water caught by the screen fall to the bottom, where they can be removed.

Another type of filter is made of a series of laminated discs placed within a large sediment bowl. This bowl acts as a settling chamber for the fuel and encloses the discs or strainer. Fuel enters the filter at the top, flows downward between the discs and then runs up a central passage to the outlet connection at the top. Dirt and water cannot pass between the discs because the clearance is too fine. Often, the carburetor carries a separate filter to clear the fuel before admission to the float chamber.

There are two common types of fuel pumps: mechanical and electrical. The mechanical pump has been common on US-built cars since 1926, but it is gradually being replaced by electric pumps. Electric pumps have long been popular in Europe.

The pump action is provided by a spring-loaded diaphragm inside a housing. An inlet valve admits fuel to the housing and blocks its return. The diaphragm exerts pressure on the fuel inside and forces it up the outlet valve, which also ensures that the fuel doesn't return into the pump. The drive is taken from an eccentric on the camshaft, which sets up a rocking motion in the pumping arm.

The action of the arm pulls down the hub of the pump diaphragm. This creates a partial vacuum which allows atmospheric pressure, acting on the surface of the fuel in the tank, to push fuel along the line and fill the pump housing. The pump arm or lever is divided in two parts at its pivot. The joint is so arranged that the part operated by the engine carries the other part around only when it is moving in a counterclockwise direction. The other part is attached to the diaphragm.

The right-hand portion of the lever is spring-loaded against the eccentric. It will continue to turn clockwise and leave behind the left-hand portion of the lever as the eccentric continues to revolve. Subsequent movement of the diaphragm is determined by

whether or not the carburetor is full of fuel. If it's full, the needle valve will prevent more fuel from entering the float chamber. The diaphragm will be kept at the bottom dead center of its travel. The right-hand portion of the lever will continue to oscillate without effect.

If the float chamber is not full, the diaphragm will be pushed upward by its spring and will force fuel into the carburetor. The left-hand portion of the lever will catch up with the right. On the next rotation of the eccentric, the diaphragm will again be pushed down in readiness for the next charging phase.

The electric pump is usually mounted away from the engine. Common locations are on the cowl or near the tank, but a few inside-tank applications exist. The diaphragm is operated electromagnetically. An iron armature attached to the diaphragm is drawn toward an electromagnet when the latter is energized. A spring tends to hold the diaphragm in the "pump empty" position. Once it reaches this position, an extension rod attached to the armature closes two contacts and allows an electric current to flow through the magnet coil. The armature is therefore pulled toward the magnet, and the retraction of the diaphragm brings fuel into the pump.

When the armature reaches the left end of its travel, the extension rod operates a throw-over mechanism which separates the contacts so that the armature and diaphragm can return to the right. The diaphragm and armature, under spring pressure, move right depending on the quantity of fuel the carburetor is ready to accept. No hand-priming device is required. The pump begins to operate as soon as the ignition is switched on.

The main advantage of the electrical pump is its immunity to vapor lock. That's a phenomenon caused by vaporization in the fuel line or pump. It means a gas is being pumped instead of fuel, with the result that the engine stalls. With electric pumps, fuel is pushed through regardless of underhood temperature.

Since 1970 in California and 1971 nationwide, cars with carbureted engines have had to be equipped with evaporation control systems. It was found that evaporation from the carburetor float bowls and fuel tanks could cause up to twenty percent of the total hydrocarbon output of a car.

These systems worked by trapping the vapors in a charcoal canister installed under the hood and then emptying the canister into the carburetor when the engine was restarted. Fuel tanks were provided with sealed caps to prevent vapor leaks. The regulations concerning evaporation control gave the industry added impetus to investigate fuel injection.

The heat value of gasoline can be directly converted into useful energy or work. Each BTU (British Thermal Unit) of heat energy can be considered equivalent to 778 ft-lb (foot-pounds) of work. One gallon of gasoline is capable of developing about 89,000,000 ft-lb of work, which is equivalent to 2,700 hp (horse-power) for one minute, or 45 hp for one hour. The higher the amount of heat energy that is converted into useful motion, the higher the engine's thermal efficiency.

Much of the fuel is wasted because of the complex process of making a car go. First, chemical energy is converted into heat energy and gas pressures. The pistons and cranks change these elements into mechanical rotation, and additional machinery finally uses this rotation to produce dynamic motion.

Present-day gasoline engines reach a thermal efficiency of about twenty-seven percent. This means that twenty-seven percent of the heat value of the fuel is converted into useful energy. The cooling water carries off about thirty percent of the heat value, and the exhaust gases blow away about thirty-three percent. Driving the cooling fan and water pump costs about three percent.

Friction losses are generally a function of engine speed. Most friction losses also vary with changes in load. An increase in piston thrust increases cylinder friction while higher-bearing loads, largely owing to distortion, increase shaft friction. On an average, friction losses account for three percent of the heat energy.

On an average, pumping losses account for four percent of the heat energy. Contrary to what you might expect, pumping losses are not proportional with engine speed. The exhaust stroke, for instance, is negative work, since there is a load on the piston crown. The intake stroke represents the bulk of the pumping losses. These losses are at maximum with the smallest throttle opening, and they diminish when the throttle is opened.

3

Combustion and thermal efficiency

The process of mixture preparation, accomplished variously by carburetors and fuel injection systems, has a vital effect on combustion and thermal efficiency. Just think of the injected engine's ability to run with higher compression ratios, and the importance of precisely controlling the air-fuel ratio becomes crystal clear. It is also an essential part of controlling exhaust emissions.

In terms of fuel savings, there is enormous promise in engines that will accept leaner mixtures. Now that every fraction of a percentage point in fuel savings is the subject of intensive laboratory investigations, the possibility of igniting and burning very lean mixtures keeps thousands of research engineers and technicians at work in the automobile industry, its supplier industries and the petroleum companies.

This avenue of research is not totally concentrated on fuel-injected engines, but the broad thrust is aimed at some system of fuel injection. Why? Because of the assumption that greater accuracy can be obtained under any and all combinations of operating conditions.

If you accept the basic truth that the energy efficiency of an engine increases as the air-fuel ratio becomes leaner and leaner, you end up with the conclusion that maximum efficiency is not reached until the engine is running on 100 percent air and not a drop of gasoline. Since 100 percent of the energy comes from the fuel, however, this conclusion cannot be correct. In practice, the engine won't run on air alone.

Just how lean an engine will run reliably and with "clean" exhaust varies continuously according to operating conditions. Consequently, the air-fuel ratio must be under continuous control with fast response to a maximum of information about temperature, speed, load and so on.

It is desirable, for fuel economy reasons, to run the engine as close to the lean limit as possible. But crossing this limit has dire consequences—it is far more dangerous than erring on the rich side.

There is risk of early flameout with overlean mixtures, which would cause an increase in hydrocarbon (HC) and carbon monoxide (CO) emissions. Leaning out the mixture even more will lead to the situation where ignition cannot occur, and the engine will stall.

For successful operation close to the limits of reliable running and meeting emission standards, the very highest precision is needed for adjusting the mixture preparation. The most attractive approach is a fuel injection system with electronic control.

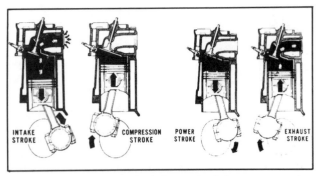

Combustion starts toward the end of the compression stroke and lasts until about two-thirds of the expansion stroke has been completed. The consistency of the charge and the timing of the spark are as important as the shape of the combustion chamber.

20

To obtain complete combustion of all the hydrogen and carbon of one pound of gasoline, we need the entire oxygen content of 14.7 pounds of air. This air/fuel ratio of 14.7:1 is called *stoichiometric,* which more or less means "ideal."

Under the right conditions, this mixture, when ignited, will release the total heat energy stored in the fuel without leaving any residual oxygen or unburned hydrocarbons. The combustion products are non-toxic, nonpolluting and harmless: carbon dioxide (CO_2) and water vapor (H_2O).

Mixtures with more than 14.7 parts of air to one part of gasoline are called *lean.* If there is less air, the mixture is called *rich.* Engineers use another ratio when they discuss rich-lean mixtures. It's the *equivalence ratio.* An equivalence ratio of one corresponds to the stoichiometric ratio.

The equivalence ratio is symbolized by the Greek letter lambda. In everyday terms, lambda can be expressed as the ratio between the actual amount of air delivered and the theoretical (stoichiometric) air requirement.

With lambda equal to 1.0, the two are identical, and the engine is receiving a stoichiometric air-fuel mixture. With lambda values less than 1.0, we are dealing with an air shortage and, therefore, a rich mixture. With lambda values higher than 1.0, we are dealing with an air surplus and, therefore, a lean mixture.

Each engine has its own lean limit, and that's why there is no fixed general number for the truly best economy figure. But it has been determined that maximum efficiency for any real engine is reached with at least half the fuel of the stoichiometric ratio, or below a lambda value of 2.0.

The car engine has another problem: Even when supplied with fuel mixture in a stoichiometric ratio, its operating parameters may not allow time for complete combustion. Consequently, even a charge that is theoretically in perfect balance will normally leave some unburned fuel in the form of raw hydrocarbons and carbon monoxide. These are expelled in the exhaust, which on most post-1975 cars is treated by a catalyst that breaks them down chemically into harmless constituents before letting them escape into the atmosphere.

Near the lean limit, hydrocarbon and carbon monoxide emissions are at their minimum. Nitrous oxide emissions are highest at the stoichiometric ratio, and fall off toward both the rich and lean ends of the curve.

The fuel injection specialist must have an acute awareness of the most intimate goings-on inside the engine if the injection system is to come anywhere close to meeting the lean-burn requirements of the engine. Thanks to the existence of advanced materials and instrumentation, research engineers can now film the combustion process in laboratory engines with

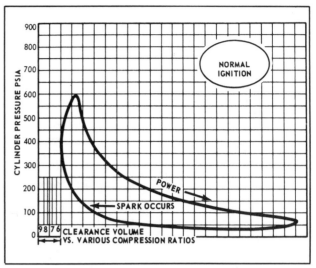

Theoretical pressure-volume diagram for a four-stroke cycle engine: The tip at lower right represents the intake stroke. Following the lower curve leftward, we see that compression brings a temperature rise. The pressure is increased as cylinder volume is diminished and sky-rockets when the mixture is ignited. Peak temperature is reached before the point of maximum pressure. Then the pressure and temperature decline during the power stroke until the exhaust valve opens. Engine design compromises resulting in heat and friction losses keep the curve theoretical—but fuel injection engines come closer than carbureted engines.

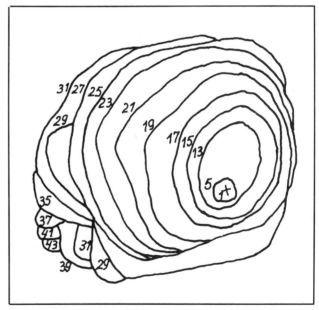

Flame front travel: The flame front spreads from the spark plug (x). By filming the combustion process through a "window" and using a mirror, it has been possible to map the progress of the flame front. This is from a Mercedes-Benz engine with parallel, canted overhead valves and fuel injection. The numbers indicate the sequence in milliseconds, revealing a critical slowing-down in the quench area adjacent to the inlet valve.

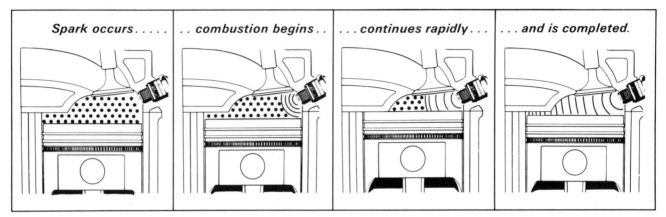

Spark occurs . . . . . | *. . combustion begins . .* | *. . . continues rapidly . . .* | *. . . and is completed.*

Normal combustion: The goings-on shown in this sequence apply equally to carbureted and fuel-injected engines.

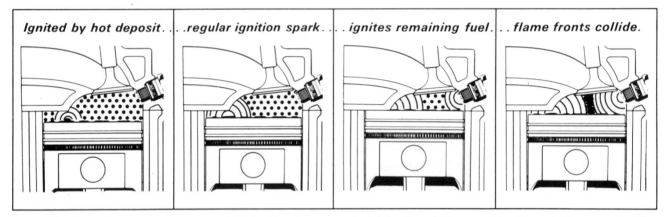

Ignited by hot deposit . . | *. . regular ignition spark . .* | *. . ignites remaining fuel .* | *. . flame fronts collide.*

These sketches show an occurrence of preignition in a typical carbureted engine. Preignition can also occur with fuel injection, being mainly dependent on combustion chamber design.

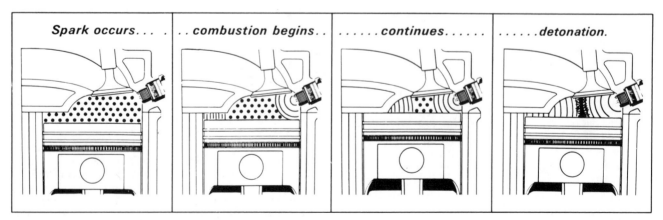

Spark occurs . . . . | *. . combustion begins . .* | *. . . . . . continues . . . . . .* | *. . . . . . detonation.*

Detonation can occur with carbureted and fuel-injected engines alike, but the risk is smaller with fuel injection. Consequently, fuel-injected engines are able to operate with higher compression ratios, which improves thermal efficiency and saves energy.

ultra-high-speed cameras, and obtain the vital data for planning, laying out, installing and setting the fuel injection system.

Normal combustion occurs when the mixture in the combustion chamber is ignited by a spark plug firing at a preset point, starting a wave of flame spreading out from the spark plug. It's like burning grass in a field. Just as it takes time for the flame to move across the field, it takes time for the flame front to travel across the combustion chamber. This flame front continues to move across the combustion chamber until it reaches the other side. The compressed mixture burns smoothly and evenly.

Flame-front speed varies from twenty feet per second to over 150 feet per second. This speed depends mainly on air-fuel ratio, compression ratio, turbulence and combustion chamber design. Flame travel is quite slow when the mixture is too rich. It is also slow when the mixture is too lean. The faster the flame front travels, the smaller the risk of abnormal combustion.

High compression gets more energy out of the fuel charge. The pressure gain in a cylinder with an 8.5:1 compression ratio is about 500 psi (pounds per square inch). The pressure gain in a cylinder with a 10:1 compression ratio is about 750 psi. It is true that more work is spent in compressing the mixture when the compression ratio is high, but there is a large gain in thermal efficiency and power because less heat is rejected to the cooling system and more of the caloric energy in the fuel is put to useful work.

Normal pressure rise is 3.5 to 4.0 times the initial pressure. Pressure-rise rates can reach 20 psi per degree of crankshaft rotation. If the initial pressure at the moment of firing is 160 psi, the peak pressure during combustion will be almost 600 psi.

Cylinder pressures are highest at wide-open throttle, and lowest during low-speed cruising or idling conditions. Under high pressure, combustion is speeded up. Turbulence in the combustion chamber will also speed up the flame front. *Turbulence,* which means a swirling airflow pattern, is designed into an engine. All types of combustion chambers can be given some amount of turbulence. Too much turbulence, on the other hand, is undesirable because it will increase the heat loss to the cylinder walls.

Conditions in the combustion chamber eventually determine the degree of success or failure of the fuel injection system. Without a thorough map of those conditions, the fuel injection specialist can do nothing for the car manufacturer.

In summary, a fuel injection system is required to do six things:

1. Give accurate control of the air/fuel ratio under all conditions, which means that it must produce the mixture that will enable the engine to reach full power at full throttle, and it must deliver a mixture that will let the engine operate with maximum thermal efficiency under part load

2. Ensure accurate control of the fuel distribution so as to provide a uniform mixture in all cylinders

3. Not permit wetting of the intake manifold walls with raw fuel, for that would upset the mixture control

4. Ensure adequate atomization of the fuel, so that a correct mixture is present in each cylinder at the moment of ignition

5. Not complicate the maintenance schedule nor introduce elements that could compromise the overall reliability or life expectancy of the car

6. Meet mass production requirements

4

Fuel injection history

Fuel injection has come a long way in the past twenty years, but its history goes right back to the early days of the carburetor. Just as the best reasons for using fuel injection today are to be found in the shortcomings of the modern carburetor, it was the lack of refinement and versatility in ancient carburetors that prepared the way for the first fuel injection experiments. The origins of fuel injection are inseparable from early carburetor history and the evolution of motor fuels.

Carburetor science began in 1795 when Robert Street achieved evaporation of turpentine and coal tar oil in an atmospherical engine (an engine working without compression of the charge). But it was not until 1824 that American inventor Samuel Morey and English patent attorney Erskine Hazard created the first carburetor (also for an atmospherical engine). Its method of functioning included preheating to promote vaporization.

A significant step on the way toward the light petroleum distillate we call gasoline was made about

The Maybach spray nozzle carburetor was refined and became the Phoenix carburetor, which inspired Arthur Krebs of Panhard & Levassor to introduce further improvements. It set the pattern for all fixed-venturi carburetors for generations.

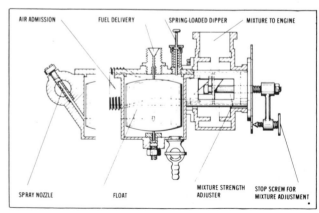

Maybach's spray nozzle carburetor introduced a fuel-feed pipe from the float chamber directly to the mixing tube, which was provided with manual adjustment for mixture strength (lean-rich).

1825. At that time, English physicist and chemist Michael Faraday was experimenting with the vaporization of liquid hydrocarbon fuels. By distilling mineral oil, he discovered benzole, which he called bicarbonate of hydrogen.

In 1833, the second and final step was taken by German chemistry professor Eilhard Mitscherlich at the University of Berlin when he succeeded in obtaining thermal splitting of benzo-acids. Out of that process came a new product—a light fuel that Mitscherlich called Benzin (gasoline in American terminology, petrol to the British).

The world was still a long way from gasoline's practical use as energy for motor power, but it came closer in 1838 when William Barnett, an English mechanic, was granted a patent for a device to vaporize gasoline. This invention was intended for use on the compression-type engine Barnett was experimenting with.

The principle of vaporization was taken forward in 1841 by an Italian scientist, Luigi de Cristoforis, who built a pistonless atmospheric engine equipped with a surface carburetor in which an airstream directed over the fuel tank was made to pick up fuel vapors.

In the years 1848 to 1850, the American doctor Alfred Drake conducted experiments with combustion engines, trying to use gasoline instead of gas. In the process, he made several types of carburetors.

In 1860, the inventor of the Deutz four-stroke gas engine, Nikolaus August Otto, began experiments with a combustion engine having a device for vaporizing liquid hydrocarbon fuels. Otto tried the engine out with white spirit. When he had no success with that type of fuel, he concentrated for a time on the development and production of gas engines.

In 1865, Siegfried Marcus of Vienna, Austria, applied for a carburetor patent. His application stressed the simplicity of his device, compared with the costly and complicated vapor generators then in existence.

In 1867, both N. A. Otto and J.J.E. Lenoir exhibited gas engines at the World's Fair in Paris. Lenoir also displayed a petroleum fuel engine with a carburetor, which was apparently overlooked at the time.

Marcus was back in 1870 with a petroleum engine that worked on the same principles as Otto's Gasmotorenfabrik Deutz engine.

The discovery of a new basic principle occurred in 1873 when Julius Hock, working in Vienna, built an atmospheric petroleum engine equipped with a primitive form of spray carburetor. The spray carburetor's superiority over the surface carburetor was not immediately apparent, however; when George Bailey Brayton began producing engines in Boston in 1874, he equipped them with surface carburetors of his own design.

In 1875, Wilhelm Maybach of Gasmotorenfabrik Deutz first converted a gas engine to run on gasoline. He was in the shop one day when he suddenly had the idea of shutting off the gas to see what would happen if he just held a rag wetted with gasoline at the mani-

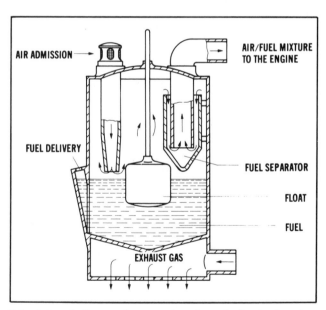

Principle of the surface carburetor: A float chamber heated by exhaust gas provided a favorable environment for evaporation. Above the fuel level, air entered through a duct controlled by a rotary valve and circulated above the surface long enough to get saturated with gasoline fumes. Then the mixture was sucked into the engine through a duct, which carried a separator that returned excess fuel to the float chamber.

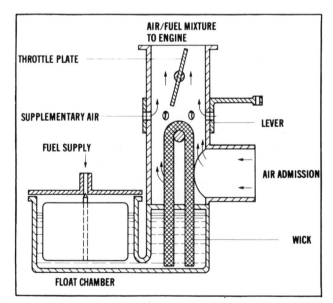

Principle of the wick carburetor: The double-ended wick, suspended on a rod, was immersed in gasoline in a chamber linked to the float chamber. Owing to the wick's porosity, fuel permeated it and the incoming air picked up fuel from its upper portion.

fold entry. The engine ran on until the rag was nearly dry.

That led to the invention of the wick carburetor. As later made popular by Frederick William Lanchester and others, this was a static type of device in which a wick soaked up the fuel on its submerged part and yielded the fuel to the air on its exposed part. Its first use in an automobile has been traced back to the motor carriage built in 1883-84 by Edouard Delamare-Deboutteville and Leon Malandin of Fontaine-le-Bourg, France.

Fernand Forest, a prolific French mechanic and inventor, invented and constructed a carburetor that included a float chamber and a fuel jet nozzle. It was fitted on a new engine Forest built in 1884.

In 1885, Otto finally obtained the results he had aimed for but failed to get in 1860, with a variety of liquid hydrocarbon fuels, including gasoline and white spirit, using an improved surface carburetor.

In the fall of 1886, Carl Benz improved his surface carburetor with the addition of a float valve to assure constant fuel level.

By 1886, Maybach had invented and tested his own type of float chamber carburetor. Finally, in 1892, he laid out the spray-jet carburetor, which became the basis for all subsequent carburetors.

Maybach never stopped searching for better ways to mix fuel and air. In 1894, he applied for a patent on a new spray carburetor, in which fuel was

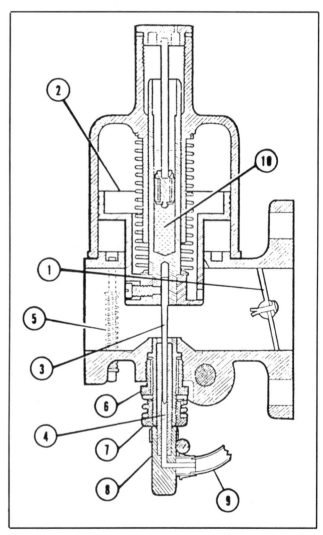

Patent sketches from the original SU patent of 1905, showing the principles of the air valve (constant-vacuum) carburetor.

Opening the throttle plate (1) in an SU carburetor causes the airflow through the venturi to speed up. Higher velocity lifts the piston (2) and tapered needle (3), which admits more fuel through the main jet (4) from the float bowl line (9). Jet size and throat diameter set limits on carburetor capacity.

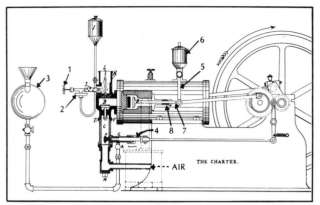

As early as 1886-87, the Charter engine combined the basic elements needed for low-pressure port-type fuel injection.

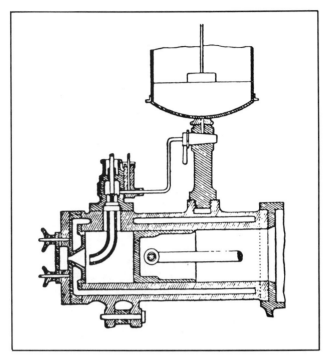

Speil's patent from 1883 shows gravity feed from the fuel tank to the port-mounted injector nozzle. Fuel metering was assured by a hand-operated linkage to the valve below the tank.

delivered in a showerhead pattern from a jet nozzle supplied from a constant-level float bowl.

The first two-barrel carburetor appeared in 1901. It was the invention of an American named Krastin, who claimed it formed consistently "good" mixtures regardless of mass airflow.

Arthur Krebs, technical director of Panhard & Levassor in Paris, invented a three-part carburetor in 1902, with automatic air bypass to minimize deviations from the ideal air/fuel ratio with increasing gas flow velocities. Krebs used manifold vacuum to open a valve and admit supplementary air.

By 1905, the carburetor had reached a basic maturity. That year, the constant-vacuum (air valve) type of carburetor was patented by George Skinner in England. The SU (Skinner Union) became popular and had many imitators. The principle is still in use today by Zenith-Stromberg as well as SU (which was taken over by Lord Nuffield about 1927 and is now part of the Rover Group).

By 1905 there also existed a body of prior art in the area of fuel injection. Its earliest entry was a patent for a compressed air metering device issued to a Frenchman named Eteve in 1881. This device fell far short of a complete fuel injection system, but it did constitute one vital element.

Another element came into being in 1883 when a German patent was issued to J. Speil for a method of injecting fresh fuel into a flame-filled chamber linked to the cylinders.

Also about 1883, Edward Butler of Erith in Kent, England, made an engine with an injection system that forced fuel under pressure through a hollow-stem inlet valve. However, Butler never developed this invention to the practical stage.

The first practical use of fuel injection was not made on an automobile, but on a stationary engine. American Franz Burger, an engineer employed by the Charter Gas Engine Company of Sterling, Illinois, developed a fuel injection that went into production about 1887. In this system, fuel was gravity-fed from the tank and entered the injector body through a throttle valve. A plunger was operated by a rocker-arm-and-pushrod mechanism from a cam mounted on a short shaft, which was gear-driven from the crankshaft. The injector nozzle protruded horizontally into the vertical (updraft) intake pipe.

In Europe, the first successful application was also made on a stationary engine, one made and set to operate at a constant speed under constant load.

The fuel-injected Deutz engine, built from 1898 to 1901, had mechanically operated valves and magnetoelectric ignition. It existed in sizes from 2 to 30 hp.

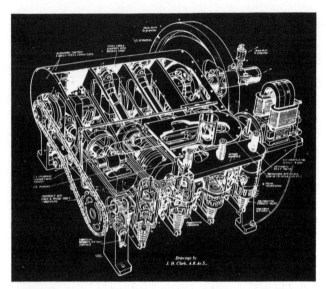

The Wright brothers used a gear-type pump to force fuel into the intake manifold of their 28 hp four-cylinder engine in 1903.

Between 1898 and 1901, Gasmotorenfabrik Deutz built 300 stationary four-stroke one-cylinder engines with low-pressure fuel injection into the intake port. The fuel was kerosene, and the injection equipment included a plunger pump with separate intake valves and pressure valves.

Use of fuel injection in airplanes also had its beginnings in America. The 28 hp four-cylinder four-stroke engine built by Wilbur and Orville Wright for their 1903 Flier was equipped with a fuel injection system. This setup used a gear-type pump, which delivered fuel under pressure into the intake ports.

As long as Wilbur Wright was alive (he died in 1912), all Flier engines were equipped with this type of fuel injection. Other aviation pioneers were quick to

grasp the reasons why the Wrights kept away from carburetors. It was not just because the airplanes needed greater freedom of maneuvering than a normal float arrangement could allow. It was also because of the risk of carburetor icing and carburetor fires, which had caused many accidents with airborne machines.

The Antoinette engine that powered Alberto Santos-Dumont's Voisin biplane, which made the first flight in Europe in 1906, was equipped with fuel injection. The brilliant Antoinette engineer Leon Levavasseur introduced not only the high-pressure plunger pump but also the principle of calibrated injectors. His injection pump was the first to have variable plunger stroke as a means of increasing or reducing the amount of fuel to be injected.

Hans Grade, an engine manufacturer in Magdeburg, Germany, began fuel injection experiments on two-stroke engines about 1905. The injection pressure was not provided by mechanical means, but by charge air precompression in the crankcase.

In 1909, a Grade monoplane powered by such an engine flew a distance of 13 km (kilometers)—the first controlled flight over German soil.

The early pioneers manufactured their own fuel injection equipment because no supplier industry stood ready to produce it. However, a leading electrical equipment manufacturer in Stuttgart soon recognized their burgeoning need.

In 1912, Bosch converted a two-stroke outboard motor to gasoline injection, using a rebuilt lube-oil pressure pump for the fuel injection. Before any

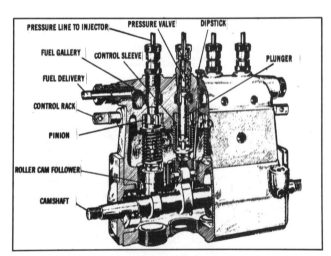

Bosch injection pump for diesel engine, made in 1927. A mechanically driven camshaft lifted a set of plungers by means of roller-type followers. The plungers entered a fuel gallery and trapped a certain amount of fuel, which was then pushed into the pressure line to the injector. Fuel volume was varied by rotating the plungers using a rack-and-pinion arrangement. This four-plunger pump became the basis for Bosch and Daimler-Benz experiments with gasoline injection.

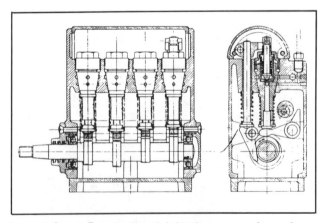

Layout for a Bosch diesel injection pump for a four-cylinder engine from 1923 shows basic cam-and-plunger action for building up hydraulic pressure. A simplified device evolved after Bosch purchased the Acro patents in 1927.

organized follow-up could be undertaken, this experiment was put in the shade when Bosch saw its priorities redirected by the demands of the Kaiser's army and navy.

Fritz Egersdörfer, an engineer with the Pallas carburetor company in Berlin, began a series of experiments with fuel injection in 1914, but he also had to put these studies aside to attend to the more urgent business of war.

By the 1920s, the carburetor industry had developed satisfactory aircraft types, and fuel injection research was put into a lengthy period of hibernation. Its chances were lessened further about 1925, when Stromberg began development work that led to the floatless injector-carburetor for aircraft engines.

During the political and military reawakening of Germany that preceded the years of Nazi rule, the DVL (Deutsche Versuchsanstalt für Luftfahrt, or German Aviation Test Establishment) had been formed at Adlershof near Berlin. It was DVL, acting under orders from the Ministry of Transport in Berlin, that brought Bosch back into research on gasoline injection. Thus began the era of the high-pressure gasoline injection system, with nozzles spraying directly into the combustion chamber.

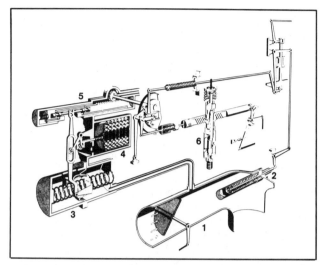

Aviation gasoline injection system for the supercharged Daimler-Benz DB 601 fighter plane engine. Direct injection was used, and the pump metered fuel to individual injectors in accordance with manifold (1) pressure and temperature (from sensor 2). The pressure signal was received by the transducer (3), whose output was modified by the aneroid bellows (4) for altitude adjustment. An amplifier (5) passed the message on to the linkage that worked the control rack. Only one plunger (6) is shown.

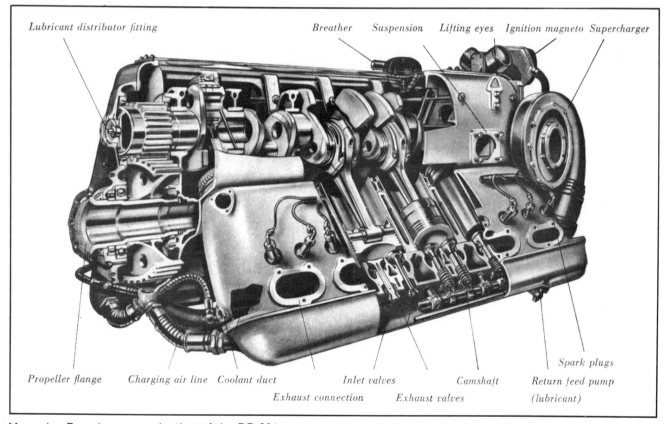

Mercedes-Benz began production of the DB 601 supercharged V-12 aircraft engine in 1937. Direct fuel injection helped lift the power-to-weight ratio to 2 hp per kg.

In 1930, Dr. Sachse of the Ministry of Transport (later with BMW) sent the DVL an order to develop an aircraft-type fuel injection system, using a BMW test cylinder. This work was carried out under the guidance of Dr. Kurt Schnauffer.

Using an ordinary diesel injection pump with eccentric shaft, plungers and spill ports, the DVL created a test engine in record time. It also began to look into the details of injector nozzle design and the control system, placing its valuable findings at the disposal of the industry.

An interim report was published on September 1, 1932, covering tests on both the one-cylinder BMW and a two-stroke DKW (Auto Union) engine. Fuel injection gave an average seven percent higher power than did carburetion on the same four-stroke BMW cylinder, with a concomitant three percent drop in specific fuel consumption. The two-stroke experiments were disappointing.

The DVL's next contract covered the conversion and testing of a six-cylinder BMW Type Va aircraft engine. It showed power gains between ten and seventeen percent on the ground-level dynamometer. Dr. Schnauffer also reported that tests were made

Constant-flow port injection was patented by Ed Winfield, developer and builder of racing car carburetors, in 1934. A gear-type pressure pump sent fuel into a rail where a pressure regulator controlled fuel flow in accordance with throttle position and manifold vacuum.

with direct injection during the compression stroke, but he was able to get more power and lower fuel consumption by injecting during the intake stroke.

After this, Mercedes-Benz—as a leading maker of aircraft engines—was pushed into fuel injection research. In autumn 1934, Mercedes-Benz began tests of a single-cylinder unit with direct injection and a Bosch diesel-type pump. At first, the standard diesel oil filter was used as a fuel filter, without a leak-oil stop. Special filters were later developed, and a leak-oil stop was added. The nozzles were also changed, from a pintle-type that sprayed fuel onto the piston crown at an angle to a multihole design.

Then the single cylinder was incorporated into a V-12 block, designated DB 601, of 33.8 liter displacement. This engine went into production in 1937 with a starting power of 1200 hp. From that point on, fuel injection conquered the aviation world.

During the 1936-39 period, Mercedes-Benz tested single cylinders from its Grand Prix racing car engines that were converted to fuel injection, both with and without supercharging. No conclusive tests were reported at the time.

At the end of World War II, it was clear to the people of Mercedes-Benz and Bosch that they possessed the key elements of a new technology. It was equally clear that no immediate application was possible. But they did not wait to start preparations (see Chapter 5).

On the Allied side, SU Carburetter Company of Birmingham, England, developed a direct fuel injection system that came into use on Rolls-Royce Merlin aircraft engines toward the end of World War II.

Simmonds Aerocessories negotiated a contract with SU for the American rights to this system, which was then used by Continental for the Patton tank engine, an air-cooled 1,790 cubic inch V-12 rated at 810 gross hp. The Patton tank came too late to see

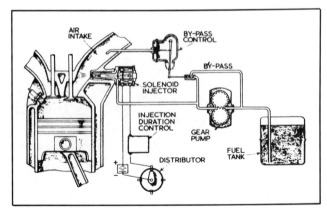

The Fuscaldo fuel injection system from 1940 worked with timed injection in each inlet port from a constant-pressure fuel line. The ignition distributor gave an engine speed reading, which was utilized to vary the duration of the injection.

action in World War II, but it was later used extensively in the Korean War of 1950-53.

By the 1950s, port-type injection was finally accepted. A patent for port-type petrol injection had been issued to engine tuner Ed Winfield in 1934, but it was more or less forgotten for many years. Another system, patented in 1935 by Ottavio Fuscaldo (designer of the OM [Officine Meccaniche] that won the first Mille Miglia in 1927), was used by Alfa Romeo on one of its entries in the 1940 Mille Miglia.

The Fuscaldo system was developed and produced by the Caproni aircraft company. It consisted of a gear-type pump that fed fuel under pressure through individual lines to each intake port. The injector nozzles carried precision-made valves which were opened by electromagnets to spray in fuel according to the needs of the engine.

In the immediate postwar world, the Winfield and Fuscaldo patents had seemed consigned to oblivion. But in 1949, a racing car appeared at Indianapolis with a fuel-injected Offenhauser engine. The injection system was invented and developed by Stuart Hilborn aided by Bill Travers. It was *indirect* injection, a straightforward, uncomplicated design not unlike Winfield's. A single throttle body at each intake port fed fuel continuously under low pressure to spray nozzles inside the port areas. This became known as *constant-flow* injection.

From 1952 to 1961, all Offenhauser-powered Indianapolis-type racing cars used Stuart Hilborn fuel injection, and Connaught adopted the system for its 1953 Grand Prix car with great success. Seeing that, European racing car constructors began asking supplier firms to develop competitive systems.

Lucas produced a successful system for the 1956 Jaguar D-Type, which won at Le Mans. This led to a production version but it was so expensive the only buyer was Maserati, for the 3500 GTi, beginning in 1961. Holley purchased the American rights to the Lucas fuel injection in 1956, but found no market for it.

The Lucas system was a port injection system with timed delivery. Fuel was pumped to a distributor at 100 psi by an electric pump. The distributor metered the fuel in accordance with the mass airflow as measured by manifold vacuum, and took care of the timing by a mechanically driven rotor with outlet ports arranged to feed fuel to each injector nozzle when the valve opened. Air intake was controlled by a sliding throttle valve linked to the accelerator.

The twin-rotor fuel distributor unit incorporated a mechanical mixture control device that responded to vacuum in the headers. Fuel quantity delivered by the pump was always in excess of the demand, the surplus being drained off and fed back to the tank.

The metering-distributor rotors were driven at half crankshaft speed from an auxiliary shaft. They contained ports that indexed with ports in an outer sleeve as well as a hollow distributing shuttle moving

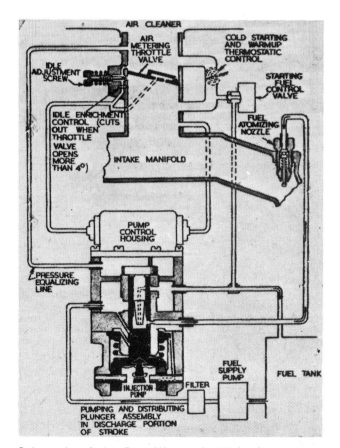

Schematic of the Borg-Warner fuel injection system, showing how all pumping action and metering functions are contained in one place: the injection pump.

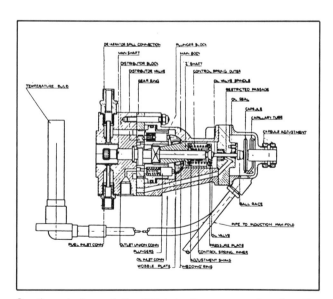

Sectioned view of the SU injection pump showing the wobble-plate operation of its plungers and the fuel-flow path through its hydraulic passages. Manifold pressure acted on a capsule with a capillary tube to affect metering.

back and forth inside the rotors. Shuttle movement was effected by means of fuel line pressure from the pump. The port alignment was designed to connect the shuttle bore with the supply line and the delivery pipe to each injector nozzle in the proper sequence.

Shuttle movement was limited by two stops: a fixed stop at the inner end and an adjustable stop at the outer end. The adjustable stop was connected to the mixture control device, which consisted of a slave cylinder with a spring-loaded piston that moved toward top dead center under high-vacuum conditions and toward bottom dead center under low-vacuum conditions. A connecting rod from the piston actuated a control wedge that lined up against two reciprocating lifters aligned with the distributing shuttles.

The stroke of the shuttle was dictated by the position of the control wedge; it was shortest when

the piston was at bottom dead center, and longest when the piston was at top dead center. By this method, the mixture was enriched during high-load conditions.

Lucas continued to develop the system, and a revised version was ready for the 1961 season BRM (British Racing Motors) Formula One racing car. Subsequent versions of the same basic Lucas injection system came into general use on Formula One engines, where it was rivaled mainly by Kugelfischer from 1966 to 1983, when Porsche went to Bosch electronic injection for the TAG (Techniques d'Avant Garde) V-6 engine fitted to the McLaren team cars.

Other suppliers also developed fuel injection systems. About 1954, American automotive engineer Ben Parsons, who had been associated with the ill-fated Tucker venture, proposed a continuous fuel injection system. Parsons attached an electrically driven variable-speed centrifugal fuel pump to the fuel tank. The electric motor was wired to a variable-output engine-driven generator and therefore "knew" at what rpm (revolutions per minute) the engine was running. It was capable of automatically matching the fuel quantity to the normal requirement, but it lacked corrective devices for air temperature, coolant temperature and altitude. Injector nozzles mounted in the intake ports were equipped to sense manifold pressure and adjust the fuel delivery according to load. Parsons did not get a fair hearing in Detroit, however, and his system was never used as original equipment on a production model.

A Bendix fuel injection unit was used on the Ford V-8-powered car that won the Indianapolis 500 in 1970. The system has four main components: mass airflow sensor, regulator, fuel-metering system and throttle, and injector nozzle. The airflow sensor works inside the main throttle body, with a combination main and boost venturi. An air diaphragm, fuel diaphragm and control valve form the regulator system. The air diaphragm is vented to the venturi airflow signal and creates a force that is opposed by the fuel diaphragm, which is vented to the fuel differential (proportional to the air differential). The fuel control or jetting system includes an idle valve to control low-end part-throttle operation and a main jet to direct fuel flow throughout the engine's power range. The nozzle system consists of pressurizing valves, which keep the two banks equalized on fuel flow to the individual nozzles.

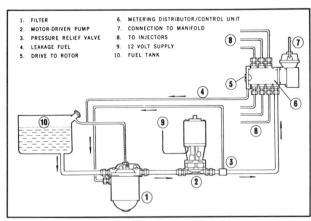

1.	FILTER	6.	METERING DISTRIBUTOR/CONTROL UNIT
2.	MOTOR-DRIVEN PUMP	7.	CONNECTION TO MANIFOLD
3.	PRESSURE RELIEF VALVE	8.	TO INJECTORS
4.	LEAKAGE FUEL	9.	12 VOLT SUPPLY
5.	DRIVE TO ROTOR	10.	FUEL TANK

Lucas fuel injection used a motor-driven fuel pump to draw fuel from the tank and pressurize it, line pressure being maintained at 100 psi by a vapor bleed relief valve. The metering distributor was driven at half engine speed from any suitable power takeoff, and it delivered metered and timed fuel charges to each injector in sequence. For production cars, fuel quantity was determined by a mixture control unit mounted as an integral part of the metering distributor, ensuring automatic response to changes in manifold vacuum. For racing cars, fuel metering was determined directly by a cam actuated by the accelerator linkage.

When Chevrolet was developing the V-8 Corvette, high-performance specialist Zora Arkus-Duntov began to look into ways of using fuel injection on it. Chevrolet engineers eventually developed a system that was put in production by Rochester Products Division (see Chapter 7). It was a port-type system, overtly based on the Stuart Hilborn design, with clever modifications and overriding controls to give the necessary flexibility for a production model sports car. Chevrolet made it optional for the 1957 Corvette.

Owing to production problems, no more than 2,750 fuel-injected engines were installed in Chevrolet and Corvette cars during 1957. Edward N. Cole, then general manager of Chevrolet, said fuel injection would again be offered in 1958 but was "too costly and complicated for general use."

Two other GM (General Motors) divisions, Oldsmobile and Pontiac, also tried the Rochester sys-

tem. The Olds engineers rejected it, but the Pontiac engineers made some modifications of their own and used it as standard for the 1957 Bonneville. Service problems in the field caused both Pontiac and Chevrolet to become disillusioned with fuel injection, and the Rochester system was no longer available when the 1959 models appeared.

No sooner had Chevrolet announced the Rochester injection system in 1957 than Borg-Warner Corporation began experimental work on a similar system. This project was assigned to the Marvel-Schebler Division and resulted in the development of a single-plunger injection pump. The pump's distribution was arranged by rotation of the plunger to index with the discharge ports to each injector line in turn. The pump gave timed injection to nozzles installed in the inlet valve ports. Nozzles were of the spring-loaded pintle type, working at a pressure of about 200 psi. The

Bendix fuel injection installation on a Ford-powered Indy car, 1970. The injection unit is mounted on the air inlet side of the turbocharger. Mass airflow measurement proved to be a good approach to holding a fixed air-fuel ratio for all accelerator positions. It was fully satisfactory in racing cars, but did not offer the versatility needed for production cars.

pump carried a control assembly whose main input was manifold absolute pressure.

This system was intensively developed in the 1960-66 period, but Borg-Warner could not sell it in Detroit. When electronic fuel injection systems began to appear, Borg-Warner withdrew from the field.

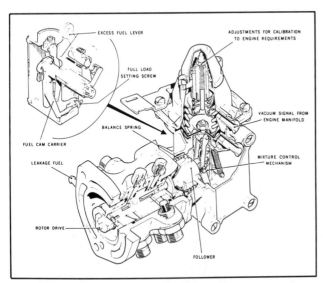

Lucas metering distributor and control unit for a multi-cylinder engine. The outlet unions for the pipes to the injectors incorporated nonreturn valves. These unions were screwed into the body and passed through the sleeve ports in rubber sealing rings. The body was an aluminum casting, and the rotor was made of hardened steel. The mixture control mechanism adjusted the shuttle travel distance in accordance with manifold vacuum or direct cam operation from the accelerator linkage.

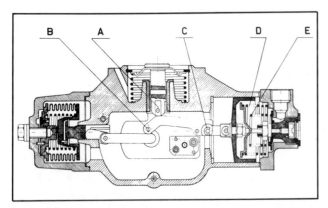

The Lucas mixture control mechanism was part mechanical, part hydraulic. Fuel line pressure was brought to bear on the central spring-loaded diaphragm, and was relayed through roller A to the follower link that carried rollers B and C. Central roller A acted on the pivoted follower, which in turn influenced the position of the fuel cam through roller B. Manifold vacuum was tapped into space D-E to control the position of the diaphragm, which then moved the pivot point for follower B-C, exerting the major influence on the position of the shuttle control stop.

American Bosch (originally a branch of Robert Bosch of Stuttgart and ultimately absorbed by United Technologies) proposed a vaguely similar system in 1957. Ford was interested, but dropped it when GM and Chrysler stopped offering fuel injection for production cars.

Like the Borg-Warner system, American Bosch's used a single-plunger pump, driven from the engine camshaft. Face cams on the spring-loaded plunger indexed with a metering sleeve in a sliding fit around the plunger, which had a fixed-length stroke. The number of cams corresponded to the number of cylinders in the engine, and the position of the meter-

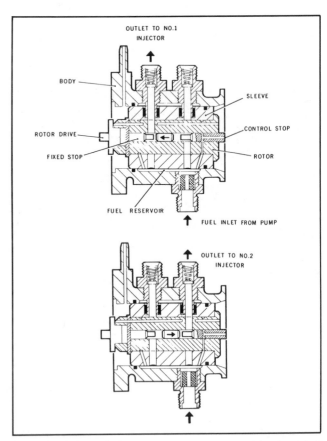

The Lucas fuel injection system relied on a principle called shuttle metering for its basic operation. For the sake of simplicity, the drawings show a two-cylinder arrangement. The engine-driven rotor has two radial ports leading to a channel in its center. The rotor fits inside a sleeve containing fuel inlet and outlet ports, and its central channel serves as a track for a shuttle valve moving axially between two stops, one fixed and the other adjustable. Rotor ports index with the sleeve ports as the rotor turns within the sleeve, simultaneously reversing shuttle movement at the proper time. Fuel pressure drives the shuttle to the fixed stop, discharging to one injector. After 180 degrees further rotation, the shuttle arrives at the adjustable stop, fueling the second injector. The amount of fuel depends on the length of shuttle travel and the bore of the channel.

ing sleeve determined the spill point, thus regulating the amount of fuel to be injected.

The face cams operated on fixed rollers carried in the drive housing, producing reciprocating motion in the plunger. The plunger had a single discharge port; on rotation of the plunger, this port lined up in sequence with the outlets to the injector nozzles. The nozzles were installed in the intake ports, and injection timing was automatically assured by plunger rotation. Fuel metering was accomplished by a control unit, which was mounted on the pump with a linkage to the main throttle body, and thus responded to manifold pressure. Commands from this unit were transmitted to the metering sleeve, which then moved axially along the plunger to blank off the spill ports as required.

A port injection system with a simple, timed plunger-type metering pump was under development by TRW (Thompson-Ramo-Wooldridge) between 1959 and 1962, but it was never released.

Friedrich Deckel of Munich, long established as a manufacturer of diesel injection equipment for farm tractors and industrial engines, developed an automotive gasoline injection system in 1960. It became known as the Kugelfischer (see Chapter 6) because the Deckel company had come under Georg Schäfer's ownership in 1955 and had been combined with his Fischer ball bearing works.

The Kugelfischer port-type injection system had timed delivery. The pump contained a cam-actuated plunger for each injector, and fuel metering was assured by an intricate regulator-cam arrangement. Air intake was controlled by a single butterfly throttle.

Tests were made with Porsche racing engines, and Peugeot became interested. After thorough test-

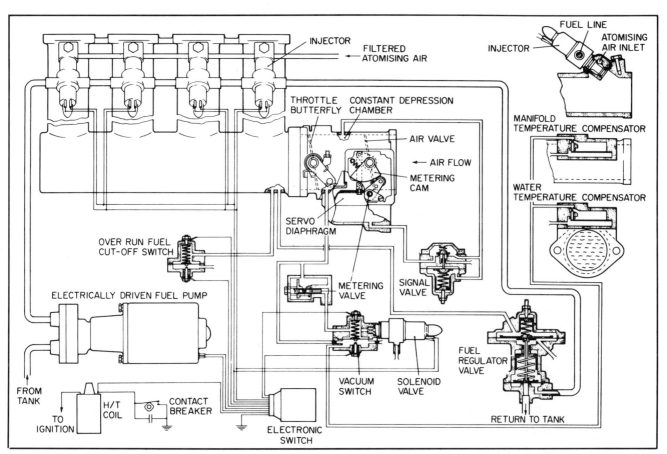

By 1974 the Tecalemit-Jackson system had evolved into this electropneumatic amalgamation with a single throttle body and airflow metering, pneumatically controlled fuel regulator valve and simple electronic switching. The constant-depression chamber, framed by a throttle valve on the downstream side and an air valve on the upstream side, was kept at a constant vacuum, and had an orifice connected to a signal valve that reacted to any pressure fluctuations. Because of the signal valve's other connection to one side of a servo diaphragm that was part of the airflow-measuring system, any drop in vacuum would bring the servo into action to close the throttle plate. The engine-driven pump of the original system had been eliminated, and fuel was sent to the injector gallery at low pressure straight from the electric fuel pump. The injectors were electrically energized by electronic switching in accordance with signals from the vacuum switch and ignition contact breaker.

ing, Peugeot standardized Kugelfischer injection on the 404 sports models in 1962, and Lancia adopted it for some Flavia types in 1965. Later versions or their successor models were fitted with more fashionable mass-produced injection systems.

The fuel injection branch of Kugelfischer was taken over by Bosch in 1974, with a small but elite production and service team that supplied BMW Motorsport for the Formula One engines from 1980 to 1985 and the Renault-Elf Grand Prix team from 1977 to 1984.

At Indianapolis Speedway, the Stuart Hilborn injection system was unchallenged until the late 1960s. Then Bendix showed up in the pits and garages with a mechanical port-type injection system originally developed for piston-type aircraft engines. Bendix called this the RS-II, and it was fitted on the turbocharged Hawk that Mario Andretti drove to victory at Indy in 1969. The following year, Al Unser's Ford-powered Indy winner also had Bendix RS-II fuel injection.

This system was mounted upstream of the turbocharger and consisted of four main components: the mass airflow meter, the regulator, the fuel control and throttle, and the nozzles.

The airflow meter consisted of a throttle body with a combined main and boost venturi, giving an air signal proportional to mass airflow. The regulator positioned a fuel control valve in accordance with the amount of air being admitted, by means of an air diaphragm that compared the mass airflow signal with the opposing force from a fuel diaphragm that was vented to the fuel pressure differential across the jetting system. This jetting system was made up of an idle valve to provide fuel for idle and part-load operation and a main jet for fuel flow throughout the power range. Air-bleed-type nozzles were inserted in the intake headers, vented to turbocharger outlet pressure.

By 1971, the Bendix RS-II system had been adopted by thirty-two out of thirty-three cars qualified to start at the Indianapolis Speedway, and it was still popular there ten years later.

A different injection system, with the cumbersome name of Tecalemit-Jackson, was used on Mario Andretti's Ford-powered Honker, prepared by John Holman and Ralph Moody for the 1967 racing season. Continuous injection was a main feature of the Tecalemit-Jackson system, which appeared in its original form in 1964. It was tested by Ford, Vauxhall, Lotus, Jaguar and Aston Martin, but it did not get approval for use on a production model at that time.

Invented by Mr. Jackson and sponsored by Tecalemit (lubricators), the system was electromechanical, with port-mounted nozzles and electronic switches. It was built up around a ring-main in which fuel circulated continuously at pressures that ranged up to 90 psi and were variable according to speed and load.

Branch pipes extending from the ring-main carried the fuel to the injector nozzles, and excess fuel was drained off and returned to the tank. The inlet manifold was fitted with two throttle valves—one connected to the accelerator linkage, and the other serving as an upstream air valve to maintain constant vacuum in the space between them. The second valve was also linked to the fuel-metering valve through a cam arrangement, and a vacuum switch determined the opening duration for the injectors.

An electric pump fed fuel to an engine-driven diaphragm pump, which increased the pressure and delivered fuel to the control unit. In the control unit, a metering valve and a sleeve regulated the amount of fuel on the basis of engine speed, manifold pressure, air density and various minor parameters.

In 1967, a subsidiary company, Petrol Injection, was formed at Plympton in Devon, England, to manufacture the Tecalemit-Jackson system. Broadspeed and others used it for racing cars with Hillman and Ford engines, and it became optional for the Lotus-Cortina in 1967.

The Tecalemit-Jackson system was still under evaluation by the motor industry as late as 1974, but lower-cost systems of proven efficiency were on the market by then, making the Jackson patents worthless.

The 1969 model Alfa Romeo 1750 for the American market used a fuel injection system developed by Spica of Livorno, Italy, a Finmeccanica subsidiary. The

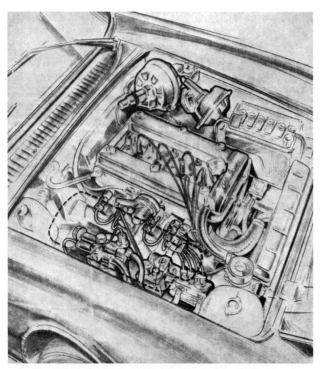

Spica fuel injection installation on the Alfa Romeo 1750. The injection pump is mounted at crankcase level, well forward, on the right of the engine.

Spica system gave timed port injection with a plunger-type pump that was basically similar to those of Bosch and Kugelfischer. The plungers had constant stroke and a normal spill-port arrangement that routed excess fuel back to the pump reservoir. Fuel metering was dependent on manifold vacuum, throttle position, barometric pressure, idle setting and coolant temperature. Sensors worked in mechanical arrangements to affect the position of a regulator cone with a cam that was free to move axially as well as rotationally, and connected to the control rack for the injection pump.

Spica fuel injection was later adapted to the 2000, the Montreal 2.5 liter V-8, the Alfetta and the Alfa Six. The equipment worked very well, but without one or two high-volume clients, Spica could not compete on cost. Production of Spica fuel injection systems was terminated in 1986.

Along with other marginal suppliers of fuel injection systems, Spica could not afford to gamble at the tables where the big stakes were made. The technology leaders with the strongest financial resources have taken the dominant roles in the fuel injection market. The evolution of fuel injection has been retarded by certain countereffects of this trend, such as the decision of some auto industry giants to produce their own systems, and the realization by some major carburetor companies that they would have no future unless they broadened their product lines to comprise a variety of fuel mixture preparation systems, with or without electronic control.

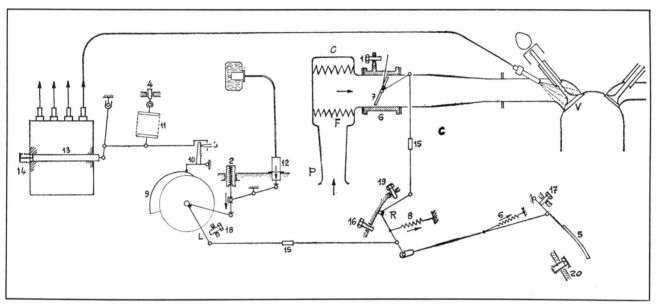

Schematic showing the Spica installation on the Alfa Romeo 1750. The injector nozzle aimed the spray directly at the back of the valve head, and the accelerator pedal was linked directly to the injection pump. 1-Throttle valve closing spring, 2-Idle speed adjustment, 3-Mixture control at idle, 4-Mixture control under load, 5-Accelerator pedal, 6-Accelerator return spring, 7-Throttle plate, 8-Accelerator linkage return spring, 9-Injection pump camshaft, 10-Control lever, 11-Aneroid bellows (barometer), 12-Coolant temperature sensor, 13-Control rack, 14-Control rack return spring, 15-Throttle valve adjustment, 16-Full-throttle stop, 17-Idle speed adjustment on pedal position, 18-Full-throttle stop on linkage, 19-Hot-engine throttle stop, 20-Full-throttle stop on pedal, C-Induction system, F-Air cleaner, G-Throttle body.

5

Bosch mechanical direct injection systems

Armed with a substantial bank of wartime experience with direct gasoline injection on water-cooled four-stroke aircraft engines, Mercedes-Benz engineers working under the direction of Max Wagner started experiments with direct fuel injection on the 1767 cc (cubic centimeter) four-cylinder engine fitted in the Type 170 V passenger car in 1946, using Bosch and Mercedes-Benz components. With its standard carburetor, this engine had a specific output of 39.5 hp per liter. Converted to fuel injection, it was made to deliver 45 hp per liter, and fuel consumption sank significantly in the low-rpm range.

The name of Mercedes-Benz is inseparable from the development and applications of the various direct

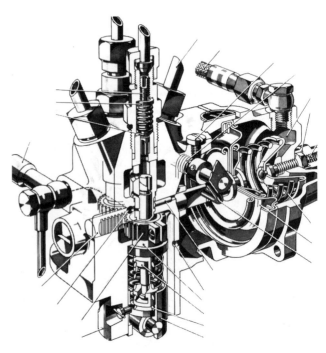

Cutaway of Bosch injection pump for two-cylinder two-stroke engines, showing the control rack with its spring-loaded diaphragm. The camshaft end is visible below the nearest plunger.

Bosch injection pump on Goliath two-cylinder two-stroke engine. The pump is driven from the nose of the crankshaft.

injection systems made by Bosch. The work done at Bosch in the early postwar period was led and supervised by two experienced engineers, W. Voit and H. Stoll.

The fuel-injected Mercedes-Benz aircraft engines were medium-speed units, governed for a maximum of 1500 rpm, having their pumps set for a full-load injection of 400 cc per stroke.

Adapting the same principles to automobile engines, which ran four times faster but needed one-twentieth the amount of fuel, was no easy task. Also, one did not need to be a marketing expert to see that the task was not urgent. In shortage-ridden postwar Germany, high-tech solutions were of little concern to people clamoring for basic transport.

Direct fuel injection and two-stroke engines

Many simple utility vehicles and primitive inexpensive cars, powered by low-cost but inefficient two-stroke engines, were aimed at satisfying this need—

and some evolved into high-volume production models. The Bosch directors put Voit and Stoll to work on fuel injection to overcome the worst drawback of the two-stroke engine: its messy gas flow control, which caused waste of fuel and loss of power.

Another disadvantage of the two-stroke engine was its lack of a proper lubrication system. Most makers simply avoided the problem by telling the users to mix a certain amount of lube oil into the fuel, which led to clogged carburetor jets as well as dirty, smelly exhaust fumes.

Two German makers of small economy cars, Gutbrod of Plochingen in Württemberg and Goliath (part of the Borgward group) in Bremen, asked Bosch for help to improve their two-stroke engines. The Bosch engineers attacked the problem in earnest in the beginning of 1949.

By 1951, Bosch had developed an injector nozzle that became the basis for all subsequent high-pressure systems. It was what the industry had been waiting for, and it was immediately adopted by both car companies. The 1951 Gutbrod Superior 700 Luxus and the

Gutbrod two-stroke two-cylinder engine with Bosch fuel injection system.

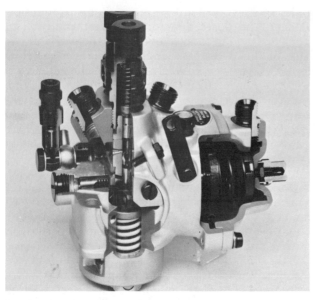

Sectioned view of the Bosch injection pump for two-cylinder two-stroke engines (PFM 2 KL 50/10).

1951 Goliath GP 700 Sport both had two-stroke, two-cylinder engines equipped with Bosch direct fuel injection.

The injector nozzles were inserted in the cylinder head adjacent to the spark plugs, with the injection timed to start at bottom dead center. The angle and shape of the spray were designed to draw maximum advantage from the natural turbulence in the two-stroke engine in order to assist the combustion process.

The injection pump was a two-plunger unit, the plungers being actuated by a cam and a lifter with a roller bearing cam follower. The effective stroke of the plunger was varied by a pneumatic governor consisting of a diaphragm connected to the inlet manifold, close behind the throttle plate.

In both cases, the lubrication system was completely separate from the fuel injection. The engine lube systems had an oil-feed pump built as an integral part of the fuel injection pump. It fed precision-metered amounts of oil directly into the intake manifold, where the oil was picked up by the air and depos-

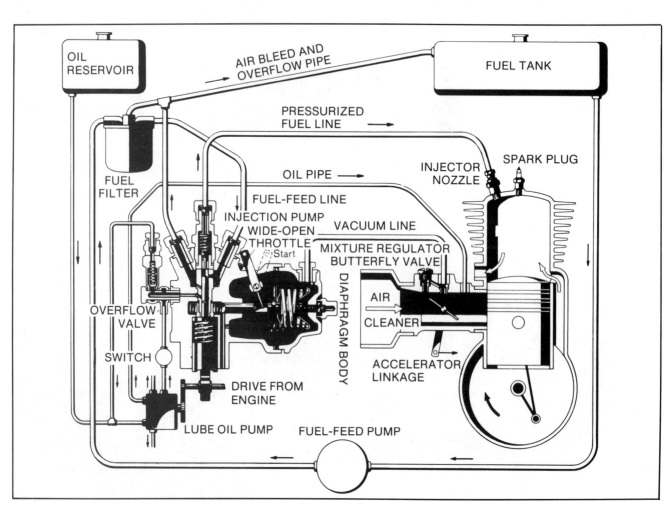

Schematic of the Bosch direct injection system for two-stroke engines.

ited inside the engine. In addition, the Gutbrod had two oil jets supplying the crankshaft bearings directly.

The Gutbrod engine, with its 663 cc displacement and 8.0:1 compression ratio, put out 30 hp at 4300 rpm, with a maximum torque of 48.1 Nm (Newton-meters) at 3500 rpm. The light (760 kg [kilogram]) front-wheel-drive vehicle had a top speed of 115 kph (kilometers per hour) and needed about forty seconds to go from 0 to 100 km/h. But the average fuel consumption was only 6.5 liters per 100 kilometers (36.15 miles per gallon), or fifteen percent less than with the carburetor version.

The Gutbrod Superior 700 Luxus burned about 220 grams per hp-hr (horsepower-hour) under full load in the rpm range around the peak-torque point, corresponding to a speed range in top gear of 70 to 100 kph. Test results below 200 grams per hp-hr were obtained by both Goliath and Gutbrod under laboratory conditions.

The Goliath was a 688 cc unit, tuned to deliver 29 hp at 4000 rpm on a 7.6:1 compression ratio. Its maximum torque was 57.8 Nm at 2400 rpm, making for better flexibility than with the Gutbrod.

The front-wheel-drive car weighed 940 kg and could reach 110 kph, taking about a minute to reach 100 kph from standstill. Its average fuel consumption was 7.5 liters per 100 kilometers (31.3 miles per gallon), or eighteen percent less than that of the Solex-equipped GP 700.

About 14,000 Gutbrod cars using this fuel-injected engine were built up to 1955, while only about 3,000 fuel-injected Goliaths were turned out before production was halted in 1957.

Direct fuel injection and four-stroke engines

Mercedes-Benz 300 SL

Experimental work on direct fuel injection on the four-stroke six-cylinder three-liter M-186 engine of the 300 series automobile began early in 1952 in the Mercedes-Benz engine laboratories, with assistance from Bosch.

The injector nozzles were copied from the two-stroke type, and an inline six-plunger high-pressure

Detail of the two-stroke injection pump installation on the Goliath two-cylinder engine.

Mercedes-Benz 300 SL in its 1952 racing edition, with triple carburetors (far side, under the hood).

injection pump was developed by Voit and Stoll, assisted by a brainy young engineer named Heinrich Knapp. The Mercedes-Benz engine adaptation was carried out by Heinz Lamm, a tester in the racing department. A 300 SL prototype with fuel injection was on the road before the end of 1952.

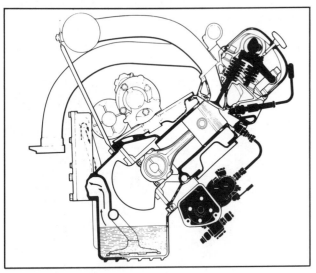

Mercedes-Benz 300 SL engine cross section, showing the injection pump (lower right) and nozzle position (inserted in the side of the cylinder block).

This was a high-performance coupe of spectacular styling, intended to do for Mercedes-Benz what the XK-120 had done for Jaguar. The 300 SL was far more radical, however, with a racing-type space frame and an advanced-design all-independent suspension system.

The reason for the interest in fuel injection was obvious: The engineers wanted to raise the compression ratio without causing detonation. They did not try to gain fuel economy by running with lean mixtures, but aimed for the fast, efficient combustion of a stoichiometric mixture while the piston was in the top-dead-center area.

The unusual shape of the M-186 combustion chamber had something to do with the engine's response to fuel injection. The lower face of the light alloy cylinder head was flat, and the chamber was formed inside the cylinder. Spark plugs were screwed into the side of the block, not the head.

On the racing prototypes, where plug changes were frequent, the plugs were nearly inaccessible, being on the underdeck side of an engine tilted fifty degrees to the left (as seen from the driver's seat). A new head—still with the upper side reserved for breathing—having plug holes as high up as possible on the lower side was produced. The plug holes in the block were filled in with blanks, except on the fuel-injected engines, where they offered a first-class location for the injector nozzles.

The 300 SL had wedge-type combustion chambers, and exhaust ports on the same side of the cylinder head as the inlet ports (not crossflow). Short runners to individual ports gave no practical ram effect. Spark plugs were almost inaccessible on the downside of the block.

Each cylinder had a fuel nozzle inserted through the top of the cylinder block, with its spray aimed directly into the combustion chamber under a pressure of about 1,100 psi.

As raced in 1952, the 300 SL prototypes were fitted with triple Solex carburetors and had 240 to 250 hp on tap at 6200 rpm. In its initial tests, the fuel-injected version put out 220 hp at a mere 5500 rpm, which hinted at a remarkable potential for improving low- and mid-range torque.

Specific fuel consumption was held in a relatively narrow band between 240 and 260 grams per hp-hr throughout the speed range for the fuel-injected version, whereas the carburetor version never got below 250 grams (except below 1750 rpm) and had peaks of 280 grams (at 5500 rpm) and 310 grams (at 3500 rpm).

At the end of the 1952 racing season, the 300 SL team was retired, and the engineers were told to develop a production model. This became available in 1954, with the engine installation angle raised to a forty-five-degree slant, plugs mounted in the head and injectors in the side of the block. With an 8.55:1 compression ratio, it delivered 215 hp at 5800 rpm and provided a peak torque of 275 Nm at 4600 rpm.

Instead of a normal intake manifold, the engine was fitted with three pairs of curved ram pipes connecting the bazookalike plenum chamber with the port flanges on the cylinder head. The ram pipes had a length of seventeen inches and a minimum internal diameter of 1.5 inches, which gave maximum ram effect between 5000 and 6000 rpm.

The 1954-55 model 300 SL had a Bosch PES 6 KL 70/320 R2 injection pump. As its designation indicates, this pump resembled the diesel type in its general layout and construction. It had a camshaft and six plungers, plunger movement being effected by the

Mercedes-Benz 300 SL engine, shown in vertical position to reveal the details of the pump drive and control system.

rotation of the camshaft. All the cams had the same contour, but they were offset relative to each other around the camshaft periphery so as to coincide with the proper firing order.

The plungers provided the pumping action that sent fuel to the injectors. When the cam lifted the plunger, the fuel held in the chamber above the plunger was forced out, into the delivery line. Fuel metering was accomplished by a spill port, which bled

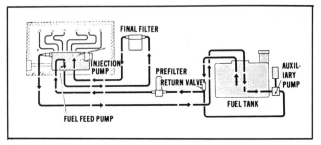

Schematic of the 300 SL injection system reveals the close relationship between this and the diesel injection system.

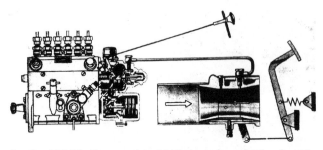

On the 300 SL, the accelerator linkage moved the throttle plate in normal fashion, while the vacuum line served to regulate the governor by means of a diaphragm.

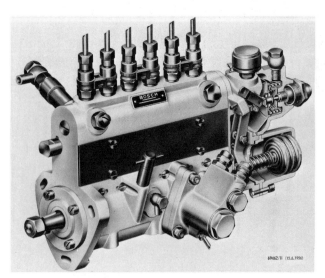

Bosch fuel injection pump, type PES 6 KL 70/320 R1, as used on the Mercedes-Benz 300 SL.

off excess fuel from each plunger when opened by a control sleeve. The sleeve had a number of slots and underwent partial rotation as directed by a rack connected to the accelerator linkage.

An improved injection pump, designated PES 6 KL 70/320 R3, was adopted in 1956. Here, the control rod for the injection pump was linked to a diaphragm placed between atmospheric pressure and manifold vacuum, thereby obtaining a measure of mass airflow. Cold-start enrichment was assured manually, from a button on the instrument panel.

The main fuel pump was mechanical. It was mounted on the injection pump and drew fuel directly from the tank through a prefilter. In addition, an electrical fuel pump mounted outside the tank fed fuel under low pressure to a T-jointed return valve inserted in the main fuel line. From the mechanical fuel pump, the fuel was pushed through a fine filter and then fed into the injection pump.

Mercedes-Benz W-196 and 300 SLR

Design work on the W-196 Grand Prix racing car began in 1952, while the team was campaigning the 300 SL coupes. The principal contenders (Ferrari, Maserati, Gordini, soon to be joined by Vanwall) were then getting about 100 hp per liter from their multi-carburetor 2500 cc engines. Fritz Nallinger, Mercedes-Benz technical director, wanted to exceed that figure by a comfortable margin. He had a lavish budget, and could afford to explore the most sophisticated technologies then available.

Nallinger's company had enough experience with fuel injection, he felt, to say that for a four-stroke unsupercharged racing engine requiring a wide speed range, a uniform mixture could best be prepared independently of the rpm and load by means of direct gasoline injection into the cylinder. With such an arrangement, he also expected that the engine would offer satisfactory torque throughout its speed range. With precise metering of the fuel quantity for each cylinder, he was sure to obtain equal cylinder pressures so as to permit loading of all the cylinders right to the limit.

Assisted by Hans Scherenberg and Ludwig Kraus, Nallinger settled for an eight-in-line layout, with the block laid on its side, thirty degrees from horizontal. With the cylinder head offset to the right side of the chassis, the inlet ducting was laid on top—with considerable ram effect—and the exhaust pipes were positioned below. The twin-cam head, with desmodromic valve operation, had a single inlet port per cylinder, aimed straight against the piston. In the center of the pentroof combustion chamber were twin spark plugs.

For breathing reasons, valve head sizes were maximized and no space was left for injector nozzles in the cylinder head. Instead, the nozzles had to be mounted in the block. Extensive experiments were necessary to determine the optimum position of the jet itself and the shape of the spray. Finally, the nozzles were placed on the inlet side of the block at an upward angle, carried in holes drilled in the cylinder barrels, and they aimed the spray directly against the exhaust valves.

The M-196 nozzles were of the pintle valve type with a single delivery hole. Fuel pressure unseated a spring-loaded one-way valve at the right moment, and no return line for leakage was needed.

On the supply side, a low-pressure pump brought fuel from the tank to a filter, and from there to the injection pump. A certain quantity of fuel returned to the tank after passing a blow-off valve. The purpose of this blow-off valve was to scavenge the inlet chamber on the injection pump and prevent vapor lock.

The Bosch injection pump was an eight-plunger design, gear-driven from a central power takeoff on the engine. Fuel delivery was accurately metered, pressurized at 1500 psi and timed to coincide with the compression stroke. There was little freedom with regard to injection timing, since the pistons shrouded

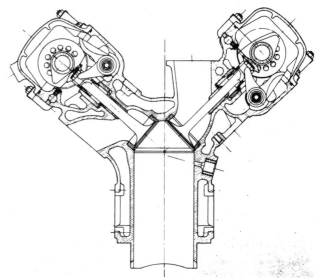

In the M-196 engine, the injector was inserted lower down in the block side, with the nozzle aimed against the center of the combustion chamber.

Mercedes-Benz M-196 engine, a 2496 cc eight-in-line unit with direct fuel injection. Bosch supplied the inline high-pressure injection pump.

Injection pump installation and drive on the M-196 Formula One engine, 1954-55.

the nozzles at the top-dead-center position. That left no more than 120 degrees of crankshaft rotation available for effective injection, which corresponds to only 0.002 seconds at 10,000 rpm.

The fuel metering was governed according to the vacuum behind a throttle plate in a venturi at the plenum chamber entry, with part-load corrections made automatically in accordance with differential pressure across the throttle plate.

When the W-196 entered its first race, the Grand Prix de l'Automobile Club de France at Reims in July 1954, the engine had a maximum output of 268 hp at 8250 rpm. By the end of the season, it was turning out 280 hp. At Monaco in May 1955, no less than 290 hp were on tap at 8700 rpm.

During the winter of 1954-55, a three-liter version (W-196 S) was prepared for installation in the 300 SLR roadsters, which won the World's Championship for Sports Cars in 1955. With a 9.0:1 compression

Mercedes-Benz 300 SLR induction system had a capacious plenum chamber and ram pipes to each intake port. The injection pump was mounted next to the intake cam-shaft. Enormous finned aluminum brake drums were mounted inboard, just ahead of the engine.

ratio, it put out 305 hp at 7400 rpm on premium gasoline.

Borgward 1500 RS

Persuaded by the Mercedes-Benz example, and by that of BMW (which adopted Bosch direct fuel injection for its four-stroke flat-twin air-cooled racing motorcycle engines), Borgward developed direct injection, with Bosch components and technical assistance, for the four-cylinder sixteen-valve four-stroke engine of its 1958 competition roadster, the 1500 RS.

The injector nozzle was mounted centrally in the pentroof-profile combustion chamber. The injection pump was driven from the rear of the inlet camshaft but was overridden for starting by an electrical pump, which was switched off manually when the engine was running.

With a 10.2:1 compression ratio, this 1.5 liter engine, based on the Isabella TS production unit, put out 140 hp. The Borgward engineers ran into material failures when higher compression ratios were tried, but by working on gas flow pulsations and fitting dual

spark plugs in each cylinder, they were able to extract 160 hp. They feared for the durability of this version, however, and restricted it to hill-climbs and sprint races.

The best-ever dynamometer reading for the 1500 RS, under laboratory conditions, was equivalent to 172 hp. Sadly, there was no sequel. Any influence the 1500 RS might have had on future Borgward products evaporated when the company fell into bankruptcy and was liquidated in 1961.

Timed port injection

In such a car as the 300 SL, the cost of the fuel injection was no problem, and owners did not object strongly to the loud noise of the high-pressure injection pump. But for the more civilized four-door Mercedes-Benz passenger cars, something quieter as well as less expensive had to be found.

Ludwig Kraus pointed out to his colleagues that it was perhaps time to divorce the injection systems they were using for spark-ignition engines with gasoline fuel from the diesel technology they had been so

With the 220 SE of 1958, Mercedes-Benz went to intermittent port injection, using a two-plunger Bosch injection pump. The two distributor boxes can be seen between the curved ram pipes.

firmly attached to all along. Because of its compression-ignition characteristic, the diesel engine demands high-pressure injection, either directly into the cylinder or into a prechamber connected to the cylinder by an open passage. That is not the case with gasoline fuel and spark ignition.

The gasoline engine doesn't care whether it breathes in fresh air or a ready-made combustible mixture. The gasoline might just as well be squirted—at low pressure—into the manifold. That might even result in more thorough atomization of the fuel droplets and thereby promote more complete combustion. And lower pressure means more than just lower noise; it also means reduced power consumption.

At Bosch, an engineer named Otto Eberle was assigned to make a study of port injection and propose a solution. As usual, Heinrich Knapp was in charge of the testing and development work.

Eberle and Knapp were stepping into unknown territory, and they had to tread carefully, always on their guard against rebellious behavior on the part of the gas, the fuel or the hardware. The thing had to work, and it had to be safe. No wonder that it ended up

being more complicated (and costly) than it really needed to be.

The initial port injection system was reserved strictly for the most expensive cars. It was first released for the 1957 model Mercedes-Benz 300 d, a long (5.19 meter) and heavy (1,950 kg) prestige limousine powered by the same three-liter six-cylinder engine that had also served as a basis for the 300 SL.

This car was named 300 d to indicate that it was fourth in the series, following the 300 a of 1951, 300 b of 1954 and 300 c of 1955. (Diesel models are identified by a capital D, as in the 180 D and 190 D.) Compression ratio was raised from 7.5 to 8.55:1, and the output climbed from 125 hp at 4500 rpm (in the 300 c) to 160 hp at 5300 rpm. Torque was boosted from 220.7 to 237.4 Nm, but the peak of the torque curve was displaced upward (the wrong direction!) from 2600 to 4200 rpm.

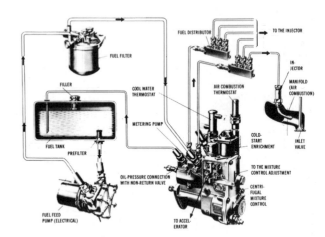

Schematic of the intermittent port injection system introduced on the 220 SE in 1958. The nozzle is separate from the injector, approaching the valve.

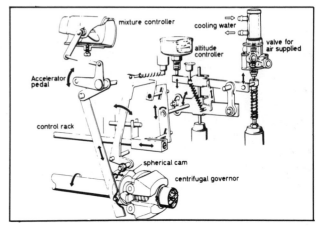

The action of the spherical cam and the two-dimensional linkage was intricate but absolutely reliable.

Second version of the two-plunger injection pump, with centrifugal regulator in place of the earlier pneumatic regulator.

Now for the technical details: This system provided metered and timed injection through six individual nozzles, which were mounted in separate runners for each port and discharged in parallel with the airflow into the ports. A six-plunger injection pump delivered fuel at 100 psi pressure to six separate lines. The stroke of each individual plunger was variable in accordance with throttle linkage position, thereby providing fuel quantities in keeping with the requirements of the moment.

The control system consisted of a venturi to measure airflow, a fuel-feed pump, a magnetic enrichment device coming into action when the coolant temperature dropped below 40°C, a diaphragm to vary fuel flow in proportion to mass airflow at any given rpm, a thermostat providing a progressive leaning-out of the mixture with rising ambient temperature and an anemometer to reduce fuel flow in high-altitude operation.

The system also included a separate electrical fuel pump, intended to ensure rapid hot-starting by eliminating vapor lock effects. The pump was started automatically when the ignition was turned on and the oil temperature was above 100°C. In addition, it could be operated manually by a separate switch to permit raw fuel to circulate throughout the system before starting the engine or to purge the circuit of vapor locks after a hot-soak.

Intermittent port injection

Mercedes-Benz

Hardly had the injection system for the 300 d been released for production than Eberle, Knapp and their assistants at Bosch began exploring every imaginable way to simplify the fuel injection system, with the

Beginning with the 230 SL in 1963, Mercedes-Benz reverted to a six-plunger injection pump, but left the injector nozzles in the ports.

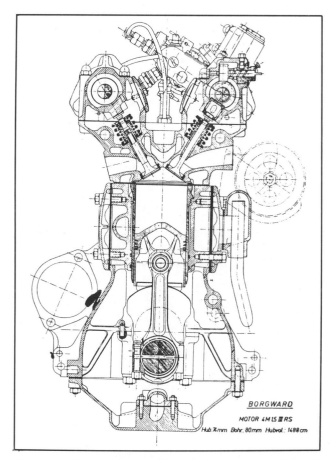

Cross section of the fuel-injected Borgward 1500 RS engine, with 16 valves in a four-cylinder engine. The Bosch four-plunger pump was mounted on the cylinder head.

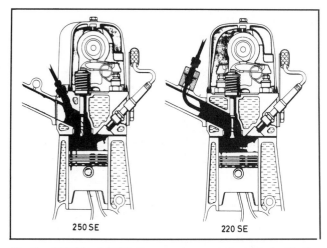

250 SE   220 SE

For the 250 SE of 1965, the nozzle was integrated with the injector, which was moved from the intake manifold to the cylinder head.

aim of making it accessible for cars produced in greater quantity.

Comparing notes on in-car testing under real-life conditions as well as laboratory test results with Mercedes-Benz engineers (who were also accumulating a lot of field experience), they came to the conclusion that while accurate fuel metering is vital, timing is not. Could they eliminate the timing function altogether? They decided to try, and found that it made no practical difference when the fuel arrived in the port area.

There are several reasons for this, starting with the frequency of delivery. Even when the engine is running as slowly as 1000 rpm, each inlet valve opens four times inside one second; at 5000 rpm, it is kicked open twenty-one times within a single second. Inside the cylinder, it is almost impossible to distinguish one pulse from another, and the overall result comes to resemble continuous injection.

That realization led to a new program: *intermittent* fuel injection. The engineers took their time, examining the concept and its full potential for simplification before doing any actual design work. One line of inquiry led to the answer that when there was no timing involved, they did not need a six-plunger pump. They almost invented single-point injection with a central throttle body, but settled for a two-plunger ZEB-series injection pump where each delivery line carried a three-branch distributor box, so as to feed all six nozzles with an *intermittent* fuel flow. One

box served the ports of the three cylinders of the front half of the block, and the second box served the ports of the three cylinders in the rear half of the block.

To simplify the fuel-metering or mixture control function, the airflow meter was eliminated and replaced by a pneumatic regulator that responded to manifold vacuum. Since manifold pressure is more or less proportional to the air quantity being drawn into the engine, it can be utilized to ensure a reasonably constant air-fuel ratio.

This system was applied to the 1958 model 220 SE, raising the output from 106 hp at 5200 rpm in the dual-Solex version of the same engine (with the same 8.7:1 compression ratio) to 115 hp at 4800 rpm. Maximum torque went up from 171.7 Nm at 3500 rpm to 186.4 Nm at 3800 rpm.

The 2195 cc six was further improved a year later, when a new-generation 220 series car stood ready. Without raising the compression, but by refining the induction system and the fuel injection hardware, the output climbed to 120 hp (still at 4800 rpm), and the torque peak was pushed up to 189.3 Nm at 3900 rpm. The system was produced up to 1965 with no further alterations.

For the 300 SE that replaced the 300 d in September 1961, a more elaborate edition of the two-plunger-pump *intermittent* port injection system was developed. As an additional refinement to assist the pneumatic regulator, fuel metering was corrected by an intake air thermostat, air pressure compensators and a coolant thermostat with supplementary air valve. An electromagnet operated the control rack for cold-starting, and the air pressure compensators served to adjust the metered amount of fuel to the ambient barometric pressure.

The 300 SE had a lightweight version of the M-186, with an aluminum block but the same cylinder dimensions. Compression ratio was 8.7:1, just as in the small-block six, and the output was 160 hp at 5000 rpm.

Minor complaints trickled in during the 300 SE's first year: lagging response, irregular starting, over-

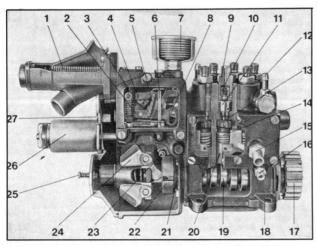

Cutaway view of the six-plunger version of the dual-row Bosch pump. 1-Temperature sensor, 2-Correction lever, 3-Connection to temperature sensor, 4-Rocker, 5-Ramp, 6-Guide bolts, 7-Altitude compensator, 8-Override guide, 9-Fuel outlet, 10-Pressure valve, 11-Pump element, 12-Fuel inlet port, 13-Rack segment, 14-Spring, 15-Lube-oil return, 16-Lube-oil inlet, 17-Pump drive cog, 18-Mounting flange, 19-Camshaft, 20-Roller follower, 21-Antithrust spring, 22-Cam face, 23-Contact roller, 24-Flyweight, 25-Idle-setting screw, 26-Stop switch, 27-Control rack access.

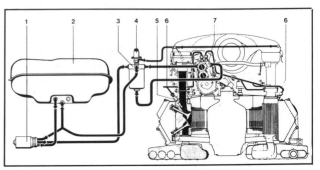

Schematic of the Bosch port injection system fitted on the 1969 models 911 E and 911 S. 1-Fuel-feed pump, 2-Fuel tank, 3-Fuel filter, 4-Cold-start solenoid, 5-Intake valve, 6-Fuel pressure line, 7-Injection pump.

sensitivity to certain conditions. Bosch engineers and Mercedes-Benz were soon hard at work on the problems. The outcome was a return to the six-plunger injection pump, with several other refinements.

The first application of the updated system was on the 230 SL, introduced in March 1963. It had a 2281 cc version of the small six, with a healthy 9.3:1 compression ratio. It put out a full 150 hp at 5500 rpm, with a peak torque of 198 Nm at 4200 rpm. For this installation, Bosch also furnished a new centrifugal mixture control regulator, which responded faster to changes in load (throttle position).

A running change was made on the 300 SE production line in January 1964, replacing the original injection system with a version of that fitted on the 230 SL. At the same time, the compression ratio went up to 8.8:1 and output climbed to 170 hp at 5400 rpm. The torque peak dipped fractionally to 249.2 Nm, but peak-torque speed rose from 3800 to 4000 rpm.

When the 220 SE was replaced by the 250 SE in 1965, the new engine (2496 cc) also had a six-plunger injection pump and the rest of the 230 SL system.

For the 250 SE engine, Mercedes-Benz went to a new type of nozzle mounting, with the nozzle inserted into a hole drilled in the cylinder head and aimed at the back of the valve head at an angle of about thirty degrees to the valve stem. The earlier nozzle type had been mounted on a manifold runner and included a long extension into the middle of the port area.

The mechanical governor of the 250 SE is also worthy of note. With the aim of obtaining better coordination among all the various operational data affecting the metering system, the injection pump control rack was linked to a roller that rested on a spherical cam. This cam was part of the governor assembly.

The connection between the spherical cam and the accelerator linkage was so arranged that changes

in throttle opening would rotate the cam about its axis. Axial movement of the spherical cam was controlled by the centrifugal governor. A high-precision unit, equipped with sensors for coolant temperature and altitude, this governor performed the duties of an engine speed sensor. It automatically adjusted air/fuel ratios for cold-starting, warm-up, cruising, coasting, sudden acceleration and changes in air density.

The roller moved up or down according to both axial and rotary motion of the spherical cam. As a result, the cam position dictated the movement of the plunger control rack. Each surface point on the cam was a reference point for an exact combination of engine load, engine speed, water temperature and altitude (air density). Engine load signals were delivered to the cam by a connection to the throttle linkage. Thus, it was the throttle opening and not the actual volume of air entering the plenum chamber that determined the air/fuel ratio.

Injection was timed to the intake stroke, but injection starting point and duration varied with engine speed. At 1000 rpm, injection began at two degrees BTC (before top center) and the nozzle stopped injecting at thirty-six degrees ATC (after top center). At 4000 rpm, the nozzle began to inject at fourteen degrees ATC and injection continued until ninety-eight degrees ATC.

The 250 SE engine put out 150 hp at 5500 rpm on a 9.3:1 compression ratio, with a peak torque of 217.5 Nm at 4200 rpm. In January 1966, this engine went into the roadster and sports coupe, which were

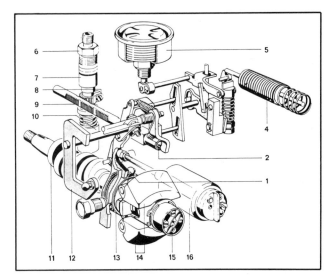

The centrifugal regulator built by Bosch for Porsche was derived from the design first used on the Mercedes-Benz 250 SE. 1-Cam roller, 2-Control rack head, 4-Temperature sensor, 5-Altitude compensator, 6-Pressure valve, 7-Plunger element, 8-Segmental pinion, 9-Control rack, 10-Roller bracket, 11-Camshaft, 12-Regulator arm, 13-Spherical cam, 14-Centrifugal flyweight, 15-Idle-setting screw, 16-Stop solenoid.

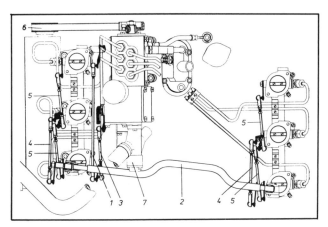

Porsche used belt drive for the injection pump on the 911 E and 911 S in 1969. 1-Pull-rod, 2-Throttle cross-connector, 3-Pump-operating rod, 4 and 5-Throttle linkage, 6-Injection pump drive, 7-Thermostat.

renamed 250 SLs. The same fuel injection system was adapted for the 280 SE and 280 SL in 1968. With 2778 cc displacement and a 9.5:1 compression ratio, its output climbed to 170 hp at 5750 rpm.

The 6.3 liter V-8 engine for the Mercedes-Benz 600 of 1964 had a similar *intermittent* injection system, with a new Bosch eight-plunger pump, centrifugal governor and nozzles mounted some distance upstream of each inlet valve. It weighed 341 kg with all its accessories and put out 250 hp at 4000 rpm. In 1968, this engine went into the 300 series vehicle, which then became a 300 SEL 6.3, a machine with exceptional performance.

Porsche

Porsche made its first experiments with Bosch direct injection on the 1500 cc flat-four in 1951, but gave up. Ten years later, it conducted some tests with the Kugelfischer system on several experimental and racing engines, but with no result for the production models. A little later, Michael May, a brilliant Swiss engineer hired as a consultant by Porsche, tried a new direct injection approach with capillary action injectors on the four-cam Carrera engine, but it never went beyond the laboratory stage.

Finally, Porsche opted for the Bosch *intermittent* port injection system for the Type 906 racing coupe that was prepared for the Targa Florio in 1966. The installation and development were carried out by Paul Hensler.

In concept, the 906 was a 911 flat-six engine in the 904 chassis. A limited-production item, sold with the Carrera 6 label, the standard 906 came with dual downdraft Webers on the 1991 cc engine, for an output of 225 hp at 8000 rpm on a 10.3:1 compression ratio.

On balance, the experience appears to have disappointed Porsche, for nothing was done during the next eighteen months to explore the use of this injection system in racing vehicles.

It was not until April 1968 that a fuel-injected Porsche 908 was entered in the 1000 km race at Monza. Though its chassis stemmed from the 904-906, the 908 was powered by a brand-new flat-eight three-liter engine. The dual-row eight-plunger injection pump, the nozzles and the control system were adapted from the Bosch components used on the Mercedes-Benz 300 SEL 6.3. But on the Porsche, the injection pump was driven by a cogged belt from the right bank inlet camshaft and mounted on top of the transmission.

With a 10.4:1 compression ratio, this ultracompact powerhouse put out 350 hp at 8400 rpm, with a peak torque of 312 Nm at 6600 rpm. For 1969, all the Porsche 908 team cars had Bosch port injection.

Then the first fuel-injected production models came out: 911 E, 911 S and 911 R. Porsche used a Bosch six-plunger dual-row injection pump fed by twin electrical fuel pumps. On its modest 8.6:1 compression ratio, the twin-carburetor 911 T boasted no more than 110 hp at 5800 rpm, and its highest torque was 157 Nm at 4200 rpm.

The 911 E (9.1:1) put out 140 hp at 6500 rpm, with a peak torque of 174.6 Nm at 4500 rpm. The 911 S (9.9:1) reached 170 hp at 6800 rpm and offered a maximum torque of 181.5 Nm at 5500 rpm. In 911 R form, with a 10.3:1 compression ratio, the two-liter flat-six put out 230 hp at 9000 rpm and gave a maximum torque of 202 Nm at 6800 rpm. Phew!

And yet, incredibly, Porsche was about to be bypassed by a technological revolution: electronics. Mercedes-Benz began switching to electronic fuel injection with the 280 E coupe in 1968. When Porsche went to a new system, it was mid-1973 and the car was the 2.7 liter 911 T. And even this wasn't electronic; it was a version of the Bosch K-Jetronic that Porsche called CIS (Continuous Injection System).

6

Kugelfischer fuel injection

The first car equipped with Kugelfischer medium-pressure port injection as a factory-installed option was the 1962 Peugeot 404 C Super Luxe. It had a pushrod ohv (overhead valve) slant-four engine of 1618 cc. In basic form, using a Solex carburetor, this engine put out 80 hp at 5600 rpm, with a peak torque of 133.3 Nm at 2500 rpm. When the Kugelfischer injection was installed, the compression ratio was raised from 8.3 to 8.8:1 and the output climbed to 96 hp at 5700 rpm, with a corresponding rise in torque to 141.2 Nm at 2800.

Next was the 1966 model Lancia Flavia 1.8, whose pushrod ohv flat-four engine normally (that is, with a two-barrel Solex) delivered 92 hp at 5200 rpm. With fuel injection but the same 9.0:1 compression ratio, it gave 102 hp, also at 5200 rpm. Peak torque increased from 147 Nm at 3000 rpm to 153.3 Nm at 3500 rpm.

Early in 1969, when the Peugeot 504 sports models replaced the 404 types, the same Kugelfischer injection system was adapted to the bigger 1.8 liter engine. This engine's displacement was raised to two

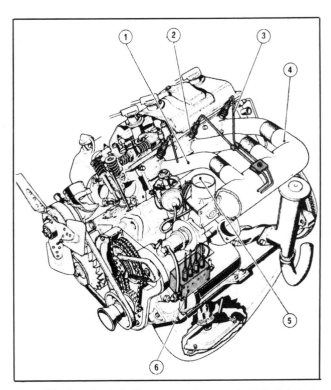

Kugelfischer installation on a Peugeot 404. 1-Intake runner, 2-Injector, 3-High-pressure fuel line, 4-Manifold, 5-Throttle body, 6-Injection pump.

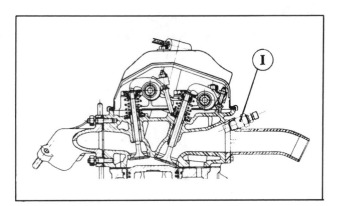

Peugeot mounted the Kugelfischer injector (1) fairly well back from the port in the crossflow head of the 404 engine.

liters for 1972. Peugeot kept the fuel-injected two-liter engine in production until October 1974, when it was replaced by the PRV V-6 engine with a carburetor. (Lancia had gone back to carburetors for the Flavia in 1970, when its engine size was increased to two liters.)

The most popular application of Kugelfischer injection was on the BMW 2000 tii, of which nearly 46,000 were produced from 1970 to 1975. The four-cylinder sohc (single overhead cam) two-liter engine

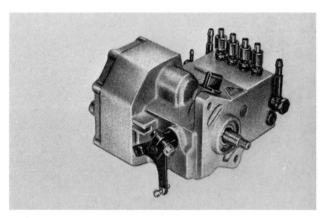

Kugelfischer injection pump for a four-cylinder engine.

put out 130 hp at 5800 rpm, giving the car a top speed of 185 kph.

In 1974, Bosch took over the fuel injection department from Kugelfischer. Since then, the Kugelfischer system has been used mainly in racing cars.

In comparison with the contemporary Bosch injection system, Kugelfischer's was perhaps less accurate. Its merit lay mainly in its ingeniously simple system of mechanical regulation.

The fuel was drawn from the tank by an electric pump and sent through a fine filter before entering the feed pump. The output line from the feed pump led to a large paper-element filter and then to the injection pump, which contained a gauze filter at the entry port.

The pump housing was a two-piece aluminum casting that contained all the main elements (a camshaft with a plunger for each cylinder). An extension of the camshaft chamber contained the gears and bearings of the control mechanism as well as the ball bearing supporting one end of the camshaft. Fuel inlet valves were mounted on the lower face of the casing, with the outlets on top.

The amount of fuel injected during each plunger stroke was regulated by limiting the return stroke of the plunger. This was done with a mechanical stop attached to a vertical lever. At one end the lever

Lancia Flavia 1.8 liter horizontally opposed four-cylinder engine, equipped with Kugelfischer fuel injection.

pivoted on an eccentrically mounted pin, while its other end made contact with a short pushrod.

The eccentric pin could be manually adjusted to deliver extra fuel for cold-starting. The pushrod, which was actuated by a cam, had the delicate job of automatically regulating the amount of fuel in accordance with engine speed and load.

The stop on the lever was located near the middle. Consequently, its position could be changed by movement at either end, so that both the pushrod and the eccentric pin could influence the length of the return stroke.

The cam that controlled the pushrod had a three-dimensional profile. Its spindle was geared to a magnetic core in a steel cylinder, which was geared to the same shaft that turned the injection pump camshaft.

Cylinder rotation around the core applied magnetic drag to the core. This drag was proportional to the speed of rotation, and the core felt it as a torque force. A gear train transmitted this torque to the regulator cam spindle. Partial rotation of the cam imparted movement to the pushrod. In addition, the cam was arranged to move axially, which also translated movement to the pushrod, in response to commands from a simple rod-and-lever connection to the accelerator linkage.

In addition to the manual cold-start button, the control lever also carried a cam-actuated stop, which set up a fast-idle mode during the warm-up period. Further, the lever that actuated the eccentric pin was connected to a thermostat that sensed the coolant temperature, which weakened the setting progressively as the engine approached normal operating temperature.

The cam was not fixed to the spindle but was merely keyed to it with a Woodruff key, which imparted rotational motion to the cam. Axial travel of

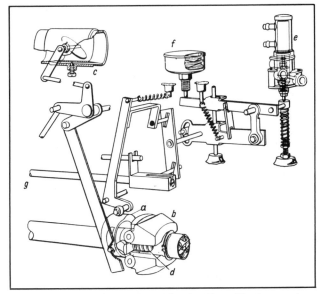

Mechanical metering control for the Kugelfischer system was built up around a cone that slid axially and turned on its axis. a-Cone, b-Centrifugal rpm sensor, c-Throttle valve, d-Adjustment springs, e-Coolant temperature sensor, f-Aneroid for altitude compensation, g-Control rod.

the cam was ensured by a peg on the end of the regulator lever from the accelerator linkage, the peg being registered with a groove machined around one end of the cam.

A circlip in a groove on the spindle kept the cam from going off the spindle at one end. To limit its travel in the opposite direction, the cam abutted the spindle bearing.

An injector nozzle was placed in each port. The main point of interest in the nozzle itself was the countercoiled spring, to prevent rotation of the needle.

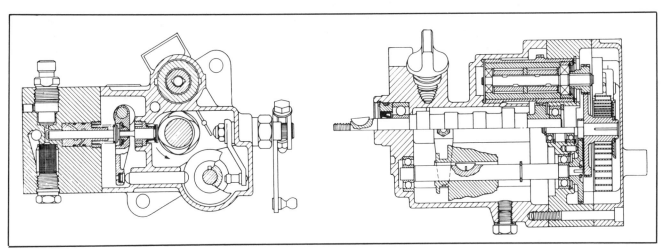

Kugelfischer injection pump belongs in the cam-and-plunger family. The gear on the camshaft drives the speed-sensing device. On the plunger's return stroke, its

movement is stopped by the regulator lever. A tappet return spring is used to keep the follower in permanent contact with the cam track.

Peugeot decided to install Kugelfischer fuel injection because of fuel economy. In tests with two-stroke outboard motors, Kugelfischer reported fuel savings up to forty percent. Other tests showed that improvements of about fifteen percent could be attained in actual driving, and that at high speeds the gains were even greater.

Kugelfischer also sold its system on the strength of its advantages in enabling engineers to raise an engine's power output. In addition to the possibility of increasing compression beyond the knock limit of the carburetor version of the same engine, the accuracy of the fuel metering ensured a torque curve of improved shape and an absence of flat spots throughout the operational speed range.

Recent racing applications, with Bosch-manufactured Kugelfischer equipment, have been on water-cooled four-stroke engines. Both Renault and BMW experimented with Kugelfischer equipment before pulling out of Grand Prix racing (BMW after the 1984 season, Renault one year later).

Renault had used Lucas fuel injection on the atmospherically aspirated two-liter V-6 used in the Le Mans racing prototypes of 1977-78, but for the turbocharged 1500 cc Grand Prix engine, it went to Kugelfischer injection. Before long, Renault engineers Bernard Dudot and Francois Castaing were getting 375 to 400 hp per liter—thereby starting the turbo era in Grand Prix racing.

BMW's racing engine specialist Paul Rosche chose Lucas fuel injection when he developed the first 1500 cc Formula Two engine with the sixteen-valve Apfelbeck head for 1967, but he tested a two-liter sports car version with eight Solex floatless carburetors. For 1970, Formula Two size went up to 1600 cc and BMW adopted Kugelfischer fuel injection.

For 1972, when the limit was raised to 2000 cc, Kugelfischer injection was retained, but BMW eliminated the redline fuel cutout. The system was pressurized at 30 psi by an electrical feed pump, and the injection pressure reached 514.5 psi. Engine response was excellent in the 6500 to 9600 rpm band, and the maximum output was 304 hp at 9250 rpm.

On the BMW turbocharged Formula One engine developed in 1980-81, the conical-surface device was discarded in favor of a mixture control system in which a spiral was moved by an electric servomotor into the exact position calculated on a continuous split-second basis by the engine's electronic control unit. By mid-1983, this turbocharged 1500 cc four-cylinder unit was putting out 580 to 600 hp at 10,500 rpm.

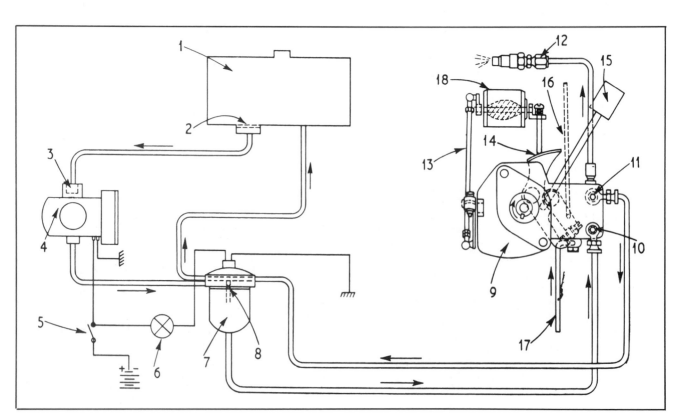

Kugelfischer system schematic. 1-Fuel tank, 2-Coarse filter, 3-Thimble-type filter, 4-Fuel-feed pump, 5-Ignition switch, 6-Warning light, 7-Main filter, 8-Air bleed, 9-Injection pump, 10-Fuel entry, 11-Excess fuel bleed-off, 12-Injectors, 13-Accelerator linkage, 14-Fast-idle cam, 15-Thermostatic control element, 16-Enriched-mixture control cable, 17-Oil pressure line, 18-Throttle body.

7

Rochester fuel injection

The widespread use of Hilborn-Travers fuel injection on Indianapolis-type racing cars no doubt influenced General Motors—and Chevrolet in particular—in its decision to develop and produce a fuel injection system for passenger cars.

By 1952, when Indy cars, almost without exception, were equipped with fuel injection, GM was no longer a newcomer in this field. Research work on fuel injection systems had been undertaken about 1948 by the GM engineering staff in liaison with Allison Division in Indianapolis, which was interested in eliminating carburetors from its aircraft engines.

As had been the case with Bosch and Mercedes-Benz, GM opted for direct injection into the combustion chamber. It undertook the conversion of a diesel engine injection pump by adding metering controls.

Within the narrow cruising-speed/full-power operating range of aircraft engines, this system did a satisfactory job of fuel distribution. The cost of a direct injection system was far too high for general use in cars, however; GM engineers estimated it at eight times that of a carburetor system.

After thorough analysis, GM engineers concluded that by injecting into the intake ports rather than the combustion chamber, the nozzle design could be greatly simplified. That would lower the system's cost and make it more attractive for automobiles.

During the initial GM engineering staff tests, it was established that no substantial power loss would be suffered by going from direct to port-type injection. There would even be advantages: While the use of individual plungers for each cylinder gave sufficiently accurate metering for aircraft engines, it tended to give poor results in a car engine at idle and in

city driving because of the erratic distribution pattern that was inherent with this system at low fuel-flow rates.

GM looked at the Fuscaldo injection system, but felt it could go further toward simplicity and obtain superior reliability. No existing fuel injection systems were considered satisfactory as a basis for further development. In consequence, GM engineers started to develop their own, guided by a few tests with single-cylinder laboratory engines.

These tests were quite thorough and proved definitively that direct injection did not offer any significant advantage over port-type injection. They also

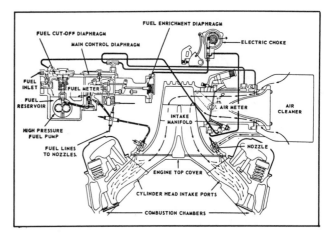

Schematic of the Rochester fuel injection system (advertised as Ramjet by Chevrolet). Despite extensive use of electrical control devices, the mechanical connections were perhaps too numerous.

57

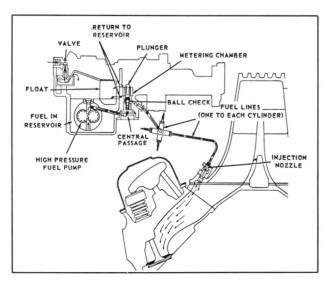

A float-controlled valve maintained the fuel level in the reservoir at a constant height, and a plunger controlled the overflow rate, with return to the reservoir.

gave proof that timing of the injection was unimportant. In fact, some tests showed that there was a power loss associated with timed injection, which also used more fuel than the rival setup with continuous injection.

Answers to the question of where to inject were clear. GM's tests showed that aiming the spray at the back of the valve head gave the most power, the lowest fuel consumption, the fastest warm-up and the best acceleration response.

From this testing, the basic principles were settled. GM was to develop a port-type injection system with continuous flow, with open-orifice nozzles directing the fuel toward the valve heads.

The basic concept as well as the detailed composition of the system took form in the mind of John Dolza. Dolza had designed the original straight-eight Buick valve-in-head engine as a youthful Fiat-trained engineer fresh from Italy in 1927. He was transferred to Allison in 1940 to work on military projects, and in 1945 moved into the power development section of GM engineering staff. Here he led the development of the axial-type swashplate refrigeration compressor from 1952 to 1954.

Dolza's assistants on the fuel injection project were Donald Stoltman, who had been responsible for the dual-carburetor setup on the 1941 Buick Super, and Ellsworth A. Kehoe, an electrical engineer who had been with Rochester Products since 1941. The adaptation to the Chevrolet V-8 engine was entrusted to Zora Arkus-Duntov, creator of the Ardun hemi-head conversion for side-valve Ford V-8 engines, who left Fairchild Aviation and in 1953 joined GM.

At the outset, Dolza hesitated between two different methods of air and fuel metering: speed-density metering, as commonly used on aircraft, versus mass airflow metering. Speed-density metering was rejected after due deliberation, mainly because it requires a means to accurately measure the engine's volumetric efficiency relative to rpm. This would be very difficult on engines working with ram-effect induction and, therefore, would be both complicated and costly.

Mass airflow metering, on the other hand, could be accomplished with a simple venturi. It was less than half as sensitive to variations in air density and temperature as the speed-density method, and fuel-density variations canceled a portion of any error.

The accelerator pedal controlled the volume of air admitted to the engine. Mass airflow was continuously metered, and a fuel meter discharged the correct amount of fuel to form an air-fuel mixture of proper lean-rich properties for the prevailing conditions.

The incoming air was measured by an annular groove as it passed through the venturi, and a vacuum signal was sent to the main control diaphragm positioned above the spill plunger in the fuel meter. The venturi was made with an annular radial entry to obtain compact construction, low frictional losses and high venturi metering signals from a relatively low pressure drop.

This system depended on two vacuum sources: manifold and venturi. Venturi vacuum was created by a diffuser cone in the air intake to the air-metering side of the manifold. The vacuum in the venturi was so slight that it had to be measured in inches of water (manifold vacuum was measured in inches of mercury, on a scale 13.6 times as large).

The control diaphragm was designed to give an accurate response to air pressure differentials of 0.01 inch of water. At the same time, it had the structural strength to withstand a thirteen-pound load, corresponding to a signal equivalent to fifty-five inches of water.

The highest pressure differential occurring in normal operation can be expressed as a ratio in excess of 3,000:1, and this differential affected the environment of both the diaphragm and the spill plunger.

The main control diaphragm was actuated by signals of venturi vacuum and directed the action of the spill plunger.

The smaller enrichment control diaphragm was actuated by the stronger manifold vacuum. It directed the adjustment of the ratio lever, which gave automatic control of mixture enrichment.

The ratio lever was controlled by a spring-loaded diaphragm subjected to manifold vacuum. It moved to the lean position at light load and to the rich position at full load, floating between the two during all transient conditions. Fuel was always present in the fuel meter, which contained a float chamber in which any vapors that formed owing to the temperature rise would be vented.

A conventional 6 psi diaphragm pump sucked the fuel from the tank, then passed it to a ten-micron primary filter and to the low-pressure section of the fuel meter. A small gear pump located below the float chamber raised the pressure and pushed the fuel through a second filter and a fuel valve to the metering cavity contained within the same housing, adjacent to the float chamber. A ball-check in the fuel valve kept the pressure sufficiently high to eliminate vapor entry into the metering cavity.

The high-pressure pump was driven from the ignition distributor shaft at the same speed as the camshaft. Under normal driving conditions, fuel pressure generally stayed below 20 psi.

Some of the fuel delivered to the metering cavity was delivered directly to the injector nozzles, while excess fuel flowed through spill ports back to the float chamber. A spill plunger, linked to the air meter, regulated the amount of fuel spill.

This spill plunger was the most vital element in the fuel metering process. It worked by balancing the force exerted on the control diaphragm (by the static pressure drop in the venturi) against the force exerted on the spill plunger (by the nozzle-metering pressure). The spill plunger was a variable bypass, constantly varying its position in response to signals from the two diaphragms.

The spill plunger was also designed to ensure that the fuel it allowed to drain back did so at normal float bowl pressure. Any excess pressurization of the spill fuel would result in fuel injection into the float bowl. That, in turn, would cause loss of the fuel through the vents and serious malfunction of the system.

In normal operation, the air velocity, acting on the throttle blade, gave a signal of manifold depression that was proportional to mass airflow. The venturi signal set up a force acting on the control diaphragm, which was transmitted to the spill plunger by means of a mechanical linkage.

Thus, increased airflow would give a stronger depression signal, putting a greater downforce on top of the spill plunger, reducing the volume of the spill flow and causing an increase in the amount of fuel delivered to the injectors.

Theoretically, a constant air-fuel ratio was maintained, because the increase in fuel flow would be proportional to the rise in mass airflow.

All the levers in the fuel-metering system were counterbalanced so that their movement was not affected by their own weight. Their positions were determined exclusively by the forces exerted by the pressure-sensitive diaphragms.

Yet the system inherently offered some degree of enrichment at idle and near-idle speeds, because the weight of the plunger and the positioning of the nozzle below the spill level in the fuel meter worked together to admit more fuel under low-volume airflow conditions than would be possible with the basic venturi system alone. This inherent enrichment was not sufficient, however, to ensure smooth idling; thus, more fuel had to be admitted from the idle orifice.

To change from power to economy, the ratio lever and its roller moved, thereby displacing the fulcrum point between the two opposing forces. The ratio lever was mounted on the same shaft as the

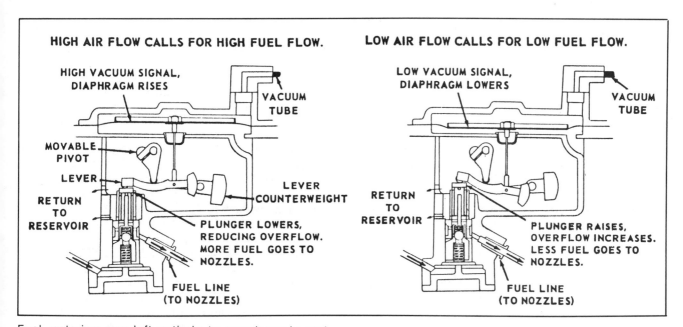

HIGH AIR FLOW CALLS FOR HIGH FUEL FLOW.

HIGH VACUUM SIGNAL, DIAPHRAGM RISES

VACUUM TUBE

MOVABLE PIVOT

LEVER

RETURN TO RESERVOIR

LEVER COUNTERWEIGHT

PLUNGER LOWERS, REDUCING OVERFLOW. MORE FUEL GOES TO NOZZLES.

FUEL LINE (TO NOZZLES)

LOW AIR FLOW CALLS FOR LOW FUEL FLOW.

LOW VACUUM SIGNAL, DIAPHRAGM LOWERS

VACUUM TUBE

RETURN TO RESERVOIR

PLUNGER RAISES, OVERFLOW INCREASES. LESS FUEL GOES TO NOZZLES.

FUEL LINE (TO NOZZLES)

Fuel metering was left entirely to aerodynamic and mechanical devices. It worked on simple principles but used a lot of components.

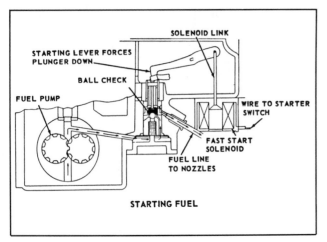

Cold-starting ability was provided by a solenoid, which opened a direct fuel passage to the nozzles. The solenoid was deenergized when the starter switch was released.

enrichment lever, and both were actuated by the spring-biased enrichment diaphragm.

To accelerate from idle speed, manifold vacuum intensified the venturi signal and the off-idle orifice began to supply extra fuel. A restriction in the bleed passage regulated the amount of fuel drained to the off-idle orifice.

Because of instantaneous response from the fuel meter, no acceleration pump was needed. When the throttle was opened, air began to rush into the manifold in amounts exceeding the stabilized air requirement, which resulted in a correspondingly stronger venturi signal.

Since the parts that made up the linkage to the fuel meter were very light and operated with minute travel, their inertia was negligible. This meant that the inrush of air would cause an immediate rise in fuel pressure, and the injectors would receive an additional shot of fuel proportional to the increase in mass airflow.

During deceleration with a closed throttle, the amount of fuel vaporization that took place in the manifold was negligible. Additionally, a coasting shut-off valve was added on top of the float chamber to avoid supplying excess fuel that would show up in the exhaust gas in the form of unburned hydrocarbons.

Under normal conditions, the coasting shutoff valve was kept closed by a spring. The valve was also linked to a diaphragm subjected to manifold vacuum. During coasting, the rise in manifold pressure vacuum would exert a force on the diaphragm. When this force exceeded the load on the spring, the valve was pulled off its seat and allowed the fuel pump to discharge directly back into the float chamber. To prevent shut-off during other conditions of high manifold pressure (such as when revving the engine free of load), the valve was connected to the accelerator linkage and could not open except on a closed throttle.

For cold-starting, a solenoid was used to open a direct fuel passage from the pump to the injector nozzles. That was done because at cranking speed, it would take twenty to thirty seconds for the fuel pump to build up enough pressure to unseat the check valve. The solenoid went into action when the starter key was turned on, and was deenergized when cranking stopped.

Gradual reduction of the cold-start enrichment during warm-up was ensured by a separate diaphragm, affected by manifold vacuum. The thermostat piston was also subject to manifold vacuum.

Under low-temperature conditions, the piston kept the ball valve below it closed, which blocked the diaphragm in full-enrichment position. During warm-up, the piston would allow the ball valve to open gradually. And when no further enrichment was needed, the full manifold vacuum came to act on the enrichment diaphragm, regulating the ratio lever position and making the piston seal off its own vacuum connection.

The injector nozzles had an accurately calibrated open orifice for the fuel flow. A small air chamber in the nozzle body, below the orifice, admitted filtered air from the air cleaner at atmospheric pressure through four holes.

This air supply made sure that the fuel discharge from the nozzle always occurred at or very close to atmospheric pressure, regardless of fluctuations in manifold vacuum. As a result, the amount of fuel injected was not subject to variables other than the metering system pressure.

The throttle was closed during idle operation, and about one quarter of the required air was admitted through the individual nozzle air chambers. The rest of the idle air bypassed the throttle blade through a separate channel, whose opening was regulated by the idle mixture screw. Therefore, the idle mixture screw determined the strength of the venturi signal caused by bleeding in manifold vacuum from the idle orifice.

Because of the long distance from the injector nozzle to the valve load, the injector body carried a tubular skirt. This skirt kept the spray within a parallel formation of 0.040 inch maximum width, which was also claimed to assist in the mixing of fuel and air. The size of the fuel orifice was determined on a basis of maximum and minimum fuel pressure needs—from an eight-inch fuel head at idle and during cranking to a maximum of 200 psi under wide-open-throttle acceleration and at top speed.

Despite all precautions and safeguards, the Rochester injection system lacked reliability in service. Moreover, Chevrolet and Pontiac engineers found that they were getting more power with carburetors (dual four-barrel and triple two-barrel setups were developed in the late 1950s). By 1959, Rochester's fuel injection system was no longer offered on new cars, remaining available only from the spare parts bin.

8

Bendix Electrojector

The world's first electronic fuel injection system for a gasoline-driven automobile engine was first described to an audience of automotive engineers at the annual meeting of the Society of Automotive Engineers (SAE) in Detroit on January 15, 1957. Robert W. Sutton revealed that since late 1952, he had been working on the problem in his laboratory at the Eclipse Machine Division of Bendix Corporation in Lockport, New York.

Sutton had filed for a patent on February 4, 1957, citing thirty-nine claims, which effectively constituted comprehensive blanket coverage for all forms of electronic fuel-injection. The patent (number 2,980,000) was granted on April 18, 1961. Bendix then secured worldwide patent coverage for this system.

The goals were to produce a system that would be easily adaptable to existing engines, with a silhouette that would permit a further lowering of hoodlines and with costs so low that it could eventually be used in mass production.

The basic invention was not a particular piece of hardware, but a complete system using fuel injector nozzles that worked by means of solenoid-controlled valves rather than by means of fuel pressure against a spring load.

The introduction of electronic controls stemmed from dissatisfaction with mechanical means of fuel metering. "Our engineers were experimenting with various fuel management systems, but we had problems getting the mechanical device to perform the same way at all engine speeds," related Sutton. "So I asked one of our engineers who was an amateur radio operator if there was any way to use electronics to help control the system. He felt there was." Sutton

reported that the engineer went out to a radio store and brought back some vacuum tubes and various other electrical components.

Sutton and his colleagues did not know it, but the use of electronic valves for delivering the fuel spray was not then a truly new idea; it was first tried in 1932 by an engineer named Kennedy who worked for the Atlas Imperial Diesel Engine Company.

At that time, Kennedy made a test installation on a six-cylinder, low-compression, spark-ignition, oil-burning marine engine. It did not have fully automatic controls or a transistor (the transistor was not invented until 1948). Still, in 1934 Kennedy installed a smaller six-cylinder engine with the same system in a truck, which was driven from Los Angeles to New

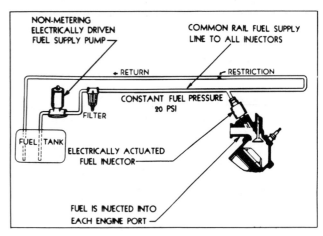

The Bendix Electrojector system featured timed injection into the intake ports, with a 20 psi common rail fuel line.

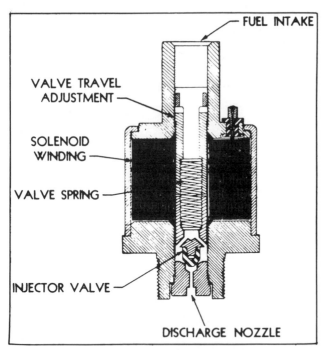

The fuel injector included a valve operated by a solenoid according to electronic signals. Fuel entered at the top end and was discharged in spray form from the nozzle at the bottom.

York and back without reliability problems of any kind. The project died when Atlas Imperial folded, and it was soon forgotten.

Thus, the Eclipse engineers started from scratch, blissfully innovating in areas where experts had stayed out from fear of running into too many difficulties. "The system we came up with was pretty primitive, but it showed promise," Sutton continued.

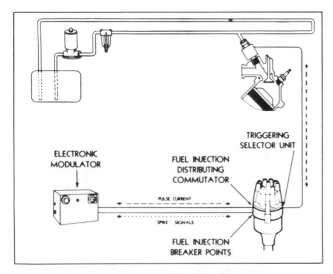

Timing pulses are sent to the injectors from the triggering selector unit (assembled as part of the normal ignition distributor). Opening duration is determined by the electronic modulator in accordance with speed and load.

"The problem was that none of us were electrical engineers, so we thought we had better consult electronics specialists from Radio Division to help us iron out some of our difficulties."

To prove that the device was workable, Sutton and two other engineers installed it on a 1953 Buick V-8 and drove it down to Towson, Maryland, with the whole electronic control unit sitting on the front floor.

"Once we got to the Radio Division, we discussed with their engineers what we had in mind," explained Sutton, "but they said there was no way to electronically meter fuel for an automobile engine. We gave them a ride in our car and afterwards they were enthused and agreed that it could possibly be done."

William L. Miron, vice president of Bendix, later recalled, "You can imagine the problems such a delicate system would have encountered in the harsh automotive environment. You can also imagine the cost of those components nearly a quarter of a century ago. Yet the basic system performed surprisingly well, if you were prepared to accept the fact that a slight problem with the electronics often resulted in an inoperative engine.

"And the cost? As one of our engineers observed not too facetiously at the time, 'We have a new electronic fuel injection system that is almost as reliable as a fourteen-dollar carburetor; and only costs twenty times as much.'"

As Sutton recalled, "We had so many problems with the vacuum tubes that we had to switch to transistors, although at the time they were very expensive.

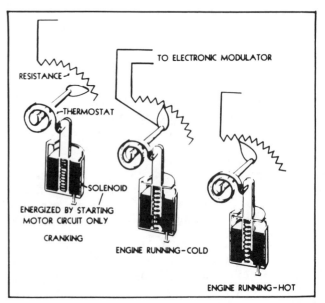

Starting enrichment—and its tapering-off during warm-up—was assured by a thermostatic control system, brought into action by the starter key. The throttle body carried a conventional fast-idle cam for warm-up conditions.

"For instance," Sutton went on, "when we used the vacuum tubes, it took thirty to forty seconds from the time we turned on the ignition for the tubes to warm up before the engine could be started. Also, sometimes when we drove under high-tension power lines the induced current would trigger the modulator to make the fuel injectors run freely and the engine would load up with fuel."

Despite the problems, Sutton obtained corporate approval for continued research on a more intensive scale. The Research Division, the Radio Division, the Friez Division and the Scintilla Division of Bendix were contacted to contribute their expertise in particular areas toward the advance of the project.

Eventually, the Eclipse engineers developed an all-transistorized control unit that became the "brain" for the Electrojector. The transistors themselves were produced at the Red Bank Division of Bendix. The unit required twelve transistors plus two silicon power transistors that cost about twelve to thirteen dollars each.

Several years later, Miron analyzed the experience: "Our engineers couldn't develop with the then-available electronics, suitable packaging to protect the system from the environment sufficiently to insure acceptable reliability. Nor could they develop an acceptable interface between the mechanical injection system and the electronic controls to assure the required levels of performance."

When the Electrojector had proved its viability, Albert H. Wickler, Jr., chief engineer for carburetors and fuel-metering systems for ground vehicles for the Bendix Corporation in South Bend, Indiana, took the project under his wing. He had joined Stromberg (which became part of Bendix soon afterwards) in 1930 and had become a leading expert not only on carburetor design, but also on manufacturing methods, tooling, quality control and cost, and carburetor adaptation to engines. By the time of his retirement in 1968, he held sixty US and foreign patents.

The basic principles behind the Electrojector were electronic control and electric actuation, with timed injection into the ports from a low-pressure (20 psi) common rail fuel system. It made use of a control system that responded to intake manifold pressure, engine rpm, ambient temperature and air pressure.

The choice of the common rail system stemmed from the basic premise that with individual lines, the effects of inertia upset the timing of fuel delivery.

Fuel was drawn from the tank by an electrically driven feed pump that maintained the fuel pressure to the fuel injector valve at 20 psi, plus or minus 0.5 psi. An ordinary fuel filter was inserted between the feed pump and the injector valves. A fine filter was not necessary, because the system had no close-fitting mechanically operating units.

The events of the operating cycle occurred in this sequence: By the cranking of the engine, the breaker point began to send out spike signals to the control unit in time with the opening of each intake valve. This triggered a multivibrator circuit in the control unit and set up a series of pulse currents tuned to a standard pulse width.

Airflow into the engine was controlled by a throttle valve of normal type but bigger dimensions. It gave

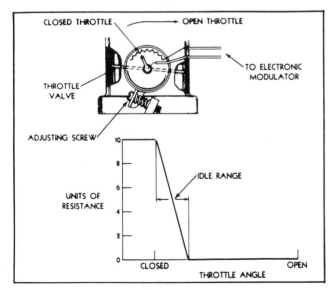

To give separate idle enrichment, a rheostat was connected to the throttle shaft. This threw a variable resistance into the control circuit whenever the throttle plate returned to its closed position.

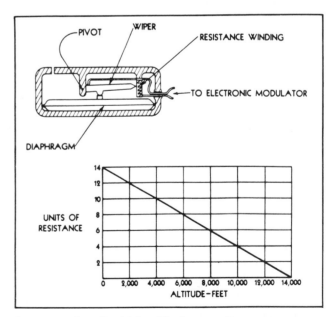

Compensation for high-altitude operation was accomplished by a device including an aneroid bellows with a variable-resistance unit that transmitted a signal to the modulator.

a fairly accurate measure of mass airflow. A pressure sensor was mounted on the intake manifold, transmitting signals about air density to the control unit.

The various sensing units worked by adding a resistance to the basic circuit, thereby modulating the pulse width of the electrical impulse to be transmitted to each injector valve in turn. Pulse widths increased as resistance in the circuits increased, which caused the injector valve to be held off the seat for an extended time, permitting additional fuel delivery.

The injectors were connected to the fuel rails and carried solenoid valves, which started and stopped the fuel delivery.

Fuel entered at the top of the injector valve, passed through the center core of the valve and discharged through the nozzle at the lower end when the valve was off its seat (open).

Bendix reported that it took a major experimental effort to develop an injector valve that could operate at high speed, maintain calibration, run with low power demand and be manufactured at a reasonable price.

The solenoid valves worked in conjunction with an engine-driven commutator. The commutator had segments shaped to vary the solenoid-energizing time in relation to engine characteristics and changes in operating conditions.

With regard to the spray formation, Bendix found that the best results were obtained when the fuel was directed at the back of the inlet valve head, so that a minimum of wetting of the manifold walls would occur.

Timing was obtained by adding a fuel injection triggering selector unit and rotor to a standard igni-

tion distributor. These elements were inserted as a sandwich between the base and the distributor cap. The triggering selector unit contained a set of breaker points and a commutator divided into sections corresponding to each injector valve.

The breaker points were actuated by the same cam that was used for the ignition system. This method ensured that for each two revolutions of the crankshaft, the fuel injection breaker points made and broke contact as many times as the engine had cylinders.

Each time the fuel injection breaker points made contact, a triggering impulse was transmitted to the electronic control unit. The modified signal was then returned to the selector element, and the impulse was forwarded to the correct injector valve.

The primary function of the electronic control unit was to transform the spike signals received from the triggering selector unit into an electrical impulse of a given standard width. It also had to coordinate engine operating signals from sensing units located on different parts of the engine and integrate these external signals into the standard pulse-width circuit, thereby modifying it in response to operating conditions.

An amplifier boosted the pulse current to the power required for energizing the injector valve solenoids. The amplified pulse was returned to the selector element and distributed to the correct injector in the same manner that a spark plug receives an ignition current. The solenoid lifted the injector valve off its seat, and fuel delivery began in the port area; from there the airflow carried the mixture into the cylinder.

As shown earlier, fuel delivery increased when the pulse width increased. Bendix pointed out that fuel delivery was not directly proportional to the pulse widths but instead followed a characteristic curve, with the strongest rise in fuel quantity accompanying the initial growth in pulse width, tapering off under continued growth.

The standard pulse segments were modified in accordance with input data from other sensors or overriding manual controls. In the prototype installation, other sensors included the manifold vacuum sensor, altitude compensator, deceleration cutoff sensor, starting enrichment control, idle mixture enrichment control and acceleration enrichment control.

The pulse width remained constant during wide-open-throttle acceleration, resulting in a fuel delivery rate proportional to engine rpm through the major portion of the speed range.

Bendix found that the fuel delivery curve would begin to dip above 4000 rpm, but that in a typical engine of that time, this dip was matched by a slowing down in the growth of air consumption, canceling the risk of air/fuel ratio disorder.

A return fuel line was part of the system, continually purging air or fuel vapor from the supply sys-

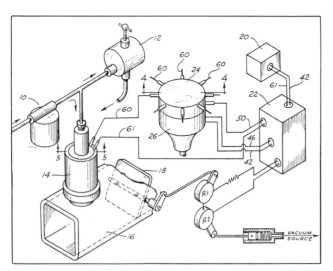

Bendix patent drawing serves as a schematic for the Electrojector, with the triggering selector unit (26) in the center. The pump (10) pressurizes the fuel, but line pressure is limited by the regulator (12) before it reaches the injector nozzle (14).

tem. Tests with high underhood temperatures and a shutdown period with empty fuel lines showed that these conditions did not lead to engine malfunctioning.

A separate idle enrichment control was provided to ensure smooth idling. It consisted of a rheostat connected to the throttle shaft, throwing a variable resistance into the control circuit when the throttle was closed.

Cold-start enrichment was accomplished by connecting a thermostat to a solenoid that could be energized only by the starter motor. The thermostat would position a variable resistance in the control circuit so that the pulse width would shrink with rising temperatures. For starting, the solenoid positioned the resistance to give increased pulse width during the cranking period. During warm-up, a fast-idle setting was obtained by a conventional (carburetor-type) fast-idle cam and thermostat mechanism.

Bendix included a provision for inserting a thermistor in the intake manifold to send signals about actual air temperature entering the engine. This signal would also be a resistance, transmitted back to the control unit to modify the pulse width and fuel delivery as required.

No acceleration pump was needed, since the manifold vacuum sensor was arranged to provide signals for mixture enrichment when required for acceleration.

A rapid change in manifold vacuum would cause the breaker points to separate, thereby introducing an additional resistance into the circuit and widening the operating pulse of the signal to the injectors long enough to equalize the pressure on both sides of the diaphragm in the sensor. In case that proved to be insufficient, Bendix also considered mechanical methods, with a connection to the accelerator linkage.

A fuel shutoff that went into action during deceleration with abnormally high manifold vacuum was provided, mainly with the motive of making emission

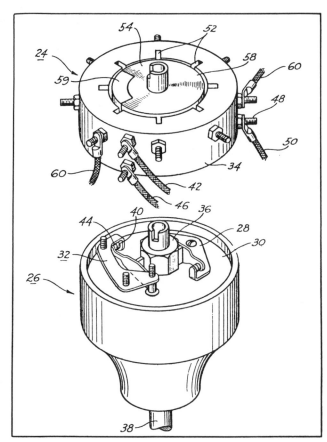

Modifications required to transform a regular ignition distributor into a triggering selector unit for fuel injection proved relatively simple. The switch (32) is mounted on a stationary contact block (34), which carries a number of output brushes (52) corresponding to the number of cylinders. Note that 50 is an input lead and 48 is an input brush.

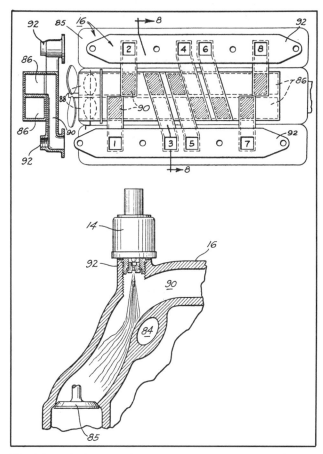

The idea of alternate fuel delivery to each bank of a V-8 engine, with simultaneous injection from four nozzles, was part of the original Bendix patents. The injector position revealed in the patent drawing may have been intended to mislead competitors, for there is no evidence that Bendix ever thought a spray directed against the manifold runner wall far back from the valve would be the most efficient method.

control easier. It also played a part in avoiding fuel waste.

By adding an aneroid bellows with a variable-resistance wire, it became possible to provide the control unit with signals about air density and thereby obtain automatic altitude compensation.

Despite high levels of refinements, the Electrojector was not successful. Why not? As Sutton explained, "The main problem with the Electrojector was that electronic components were so expensive that the system was priced right out of the market. Also, people at that time were more concerned about horsepower than fuel economy and emission control."

Without original equipment contracts, Bendix put the project on the back burner in 1960. Miron recounted: "One of my first assignments when I came to Bendix in 1961 was to kill our electronic fuel injection program. At that time, Bendix had already invested more than $1 million in a system that appeared to hold little promise as a viable product for the corporation because of the poor cost-benefit ratio and its apparent inability to survive in the harsh automotive environment."

Miron continued: "So we shelved the program and it wasn't dusted off until several years later when two major developments occurred: first, the demand for a more accurate method of metering fuel in order to reduce exhaust pollutants; and second, technological advances which made an electronic fuel control system more economically feasible. These technical advances included the development of wire harnesses and connectors capable of transmitting electronic signals at the critical junctures, suitably rugged circuit designs and circuitry, mechanical mountings and interfaces that could survive the automotive environment—a host of interrelated problems had been resolved."

The renewed effort was to awaken Cadillac's interest and lead to Cadillac's use of Bendix electronic fuel injection on the 1976 model Seville.

In 1966 and 1968, Bendix had granted patent licenses to Bosch for electronic fuel injection manufacturing rights in Germany and Brazil, and for world sales except for Canada and the United States.

In 1968, Bendix and Bosch also signed a mutual technical assistance and cross-license agreement. Then in 1969, Bendix and Bosch jointly negotiated license agreements with Nippondenso, Japan Electronic Control Systems Company. Two years later they signed a similar contract with Joseph Lucas Industries. In 1974, the Bendix-Bosch cross-license and technical assistance agreement was broadened to include zirconium oxide exhaust gas sensor technology.

9

The electronics revolution

Although the Bendix Electrojector failed to meet its maker's cost objectives and gain acceptance in Detroit, the force of the coming electronics revolution was growing. Inevitably, electronics would come to dominate the fuel injection picture and revolutionize fuel injection technology, just as it had revolutionized ignition systems and instrumentation.

That was not evident, however, in 1960. Only the most ardent believers were then spending time and money on such projects.

Electronic fuel injection

The principle of electronic fuel injection is very simple. The injectors are opened not by the pressure of the fuel in the delivery lines, but by solenoids operated by an electronic control unit. Since the fuel has no resistance to overcome, other than insignificant friction losses, the pump pressure can be set at very low values, consistent with the limits of obtaining full atomization with the type of injectors used.

The amount of fuel to be injected is calculated by the control unit on the basis of information fed into it regarding the engine's operating conditions. This information includes manifold pressure, accelerator enrichment, cold-start requirements, idling conditions, ambient temperature and barometric pressure. The systems work with constant pressure and with variable timed or continuous flow injection.

Compared with mechanical injection systems, the electronic fuel injection has an impressive set of advantages. It has fewer moving parts, no need for ultraprecise machining standards, quieter operation, less power loss, a low electrical requirement, no need

for special pump drives, no critical fuel filtration requirements, no surges or pulsations in the fuel line and finally, the clincher for many car makers, lower cost. Unfortunately, its price is still too high compared with that of a carburetor.

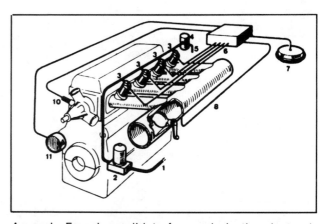

An early French candidate for a role in the electronic revolution was the Sopromi-Monpetit system. A mechanical gear-type fuel pump pressurized the fuel and fed it to a fuel rail through a pressure regulator that limited the pressure to a constant 220 to 250 psi, with a return line. The electronic control unit received signals with information on atmospheric pressure, manifold vacuum, coolant temperature and engine speed. Solenoid valves integrated with the injectors gave timed injection with metered amounts by staying open for longer or shorter durations. 1-Fuel from tank, 2-Fuel pump, 3-Injector, 4-Pressure regulator, 5-Return line, 6-Control unit, 7-Atmospheric pressure sensor, 8-Manifold vacuum sensor, 9 and 10-Engine speed sensor, 11-Coolant temperature sensor.

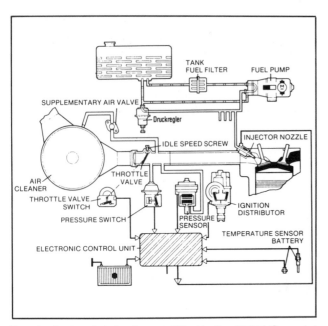

Bosch electronic injection as fitted to the 1968 US-model VW 1600.

Even if Bendix could not make any headway in Detroit, the British group Associated Engineering (AE) felt that Europe's industrial fabric as well as its market created conditions different from those in America. In 1966, the company decided to offer its Brico electronic fuel injection to selected producers of high-performance and prestige cars.

After two years of testing, Aston Martin made the Brico system optional on the 1969 model DB 6 Mark II. Power and torque were unchanged from those for the Vantage engine with the same 9.4:1 compression ratio and triple two-barrel Weber carburetors: 330 hp at 5750 rpm, 397.3 Nm of torque at 4500 rpm. Though few customers bought the fuel-injected version, it was offered as long as the six-cylinder four-liter engine remained in production (up to 1975).

The AE Brico system featured timed fuel delivery into the ports with solenoid-operated valves whose opening duration—dictated by electric pulses from the control unit—determined the amount of fuel to be injected. The electronic control unit was made up in two sections, a pulse generator on one side and a

The AE Brico was one of the first electronic systems, being offered to the industry as early as 1966. It fell by the wayside, however, its only application being in the Aston Martin DB6 Mark II of 1969. The AE Brico offered timed and metered fuel delivery to individual injectors from a constant-pressure ring-main.

combined computer and discriminator on the other, using transistors and printed circuits.

Fuel-metering calculations were based on measurements of manifold absolute pressure and combustion air temperature. The injector nozzles were connected in parallel across a ring-main and injected a conical-pattern spray into the inlet ports. An engine-driven fuel-feed pump delivered fuel to a float chamber. This chamber ensured a constant supply for the electrically driven high-pressure pump that kept the fuel circulating in the ring-main under 25 psi pressure.

The French auto industry looked at, and turned thumbs down on, a rival system invented by Louis Monpetit. Covered by a series of patents issued between 1967 and 1973, Monpetit's Sopromi system used an electronic control unit for actuating solenoids mounted on port-type injectors. The control unit received signals from a mechanical rpm sensor mounted on the camshaft, a coolant temperature sensor mounted on the water pump housing, a manifold vacuum sensor and a barometer to indicate atmospheric pressure.

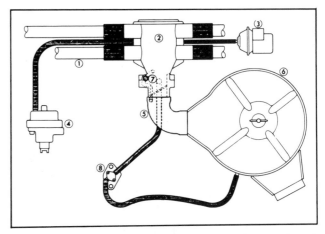

1968 US-model VW 1600. At idle, the throttle was closed, and the engine ran on bypass air. 1-Intake manifold, 2-Intake air distributor, 3-Pressure switch, 4-Pressure sensor, 5-Throttle valve body, 6-Oilbath air cleaner, 7-Idle air adjusting screw, 8-Air bleed valve.

Bosch engineers and executives were not blind to what was going on. Without fanfare, they secured world rights to the Bendix patents for electronic fuel injection in 1965.

This act was in part prompted by pressure from Volkswagen (VW). Fearful that the new antipollution laws in America could result in its carburetor-equipped cars being barred from its biggest export market, Volkswagen asked Bosch to come up with a solution. The experimental departments of both companies tackled the problem. A program led by Hermann Scholl on the Bosch side and Werner Buttgereit in Wolfsburg adapted the Bendix electronic fuel injection principles to the air-cooled flat-four VW 1600 (Type 3) engine in record time.

Since the objective had been to reduce exhaust emissions and not to raise the power output, horse-

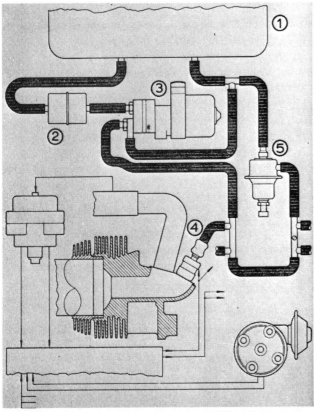

A rotating roller-vane pump in the 1968 VW 1600 installation fed fuel to a pressurized rail that supplied all injector valves. A pressure relief valve was set to open at 28 psi, letting excess fuel run back to the tank. 1-Fuel tank, 2-Fuel filter, 3-Electric fuel pump, 4-Solenoid injection valve, 5-Pressure relief valve.

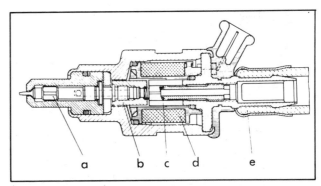

The Bosch Jetronic injector contains a solenoid valve, the needle serving as an armature and the magnetic windings being carried in the valve body. Pulling the needle off its seat to about 0.006 inch clearance opens the flow to the nozzle. Opening and closing times are about one millisecond. a-Nozzle needle, b-Armature, c-Coil spring, d-Magnetic windings, e-Fuel entry.

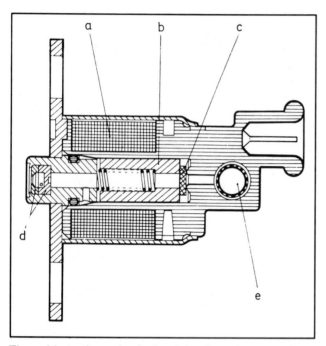

The cold-starting valve in the Jetronic system contains a simple solenoid valve, spring-loaded in its closed position. No rapid-action demands are placed on this valve, which is energized by the starter motor if the temperatures are low enough to let the control unit open the circuit to the valve. a-Magnetic windings, b-Armature, c-Seal, d-Nozzle, e-Fuel line.

power remained the same (54 DIN at 4000 rpm). This version went into production in the summer of 1967 for the US-market 1968 model Volkswagens.

The actual hardware corresponded in broad strokes to that of the analog computer, timed port injection system Bendix was then concocting in America.

The VW/Bosch Jetronic system relied on manifold pressure sensors for air metering and load measurement, pressurized fuel running in rails and metered by solenoid valves at each injector, and excess fuel flowing back to the tank through a return line.

The main parameters used for calculating the amount of fuel were engine speed and manifold absolute pressure. Since the fuel line pressure was constant, the metering rate through the nonvariable discharge port of the injectors was constant. For more fuel, the injectors were simply kept open for a greater length of time.

Engine speed data were fed continuously to the control unit from a pulse generator in the ignition distributor. This little device also measured the angular position of the crankshaft, which determined the timing of the injection (fifteen degrees after top dead center).

Fuel supply to the pressure line was ensured by a rotary roller-vane pump, which was fitted with a relief valve to limit the pressure to 28 psi. The electrically driven pump was activated by the ignition key

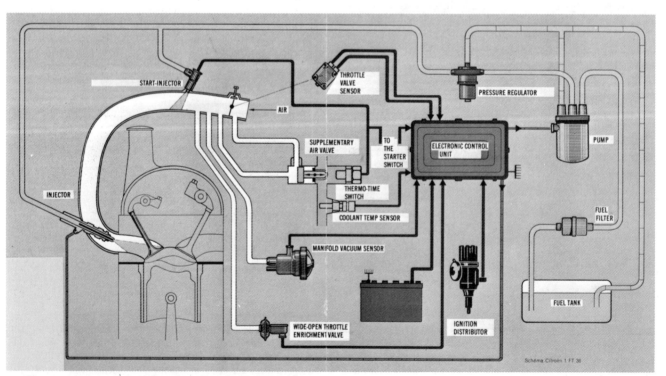

The Bosch Jetronic installation on the 1970 Citroen DS-21 had a separate cold-start injector mounted at the manifold entry.

plus (1) starter motor operation or (2) for push-starting, an engine speed of at least 200 rpm.

After passing through an oil-bath air cleaner, the fresh air was ducted through a throttle body to an induction air distributor, feeding into a four-branch inlet manifold. Air pressure in the intake duct was monitored by an inductive pressure sensor having no mechanical parts, which was inserted in the distributor section.

Five correction systems were programmed into the control unit. These ensured a richer mixture for all starts, a richer mixture for cold-starts, enrichment during warm-up, enrichment during wide-open-throttle operation and fuel shutoff on deceleration (above 1000 rpm).

For cold-starting and warm-up, an air bleed bypass valve, operated by a bimetal temperature-sensing element inside the crankcase, opened to admit a supplementary amount of fuel at idle. Above 10°C the crankcase sensor signals were ignored, while a secondary temperature sensor in the cylinder head provided data for the control unit.

This system lacked altitude correction, since tests showed that manifold absolute pressure automatically responded to variations in barometric pressure.

Bosch lost no time in developing the Jetronic system into a general package that was suitable for water-cooled as well as air-cooled four-stroke engines.

Identified by some users as the D-Jetronic, it was fitted on the 1969 model Opel Admiral, Citroen DS 21, Volvo P-1800 E, Saab 99 E and Mercedes-Benz 250 CE, plus all Mercedes-Benz models powered by the 3.5 and 4.5 liter V-8 engines. These were followed by the Lancia Flavia in 1970 and the Renault 17 in 1971.

Electronically controlled carburetors

Worried about losing its markets to fuel injection, the worldwide carburetor industry eagerly seized on electronics as an important weapon in the defense of its own basic product. The same sensors and control system used for fuel metering in electronic injection systems were directly applicable to carburetors.

As early as 1975, leading carburetor manufacturers such as Weber, Solex, Zenith, Stromberg, Carter and Holley were experimenting with electronically controlled carburetors. Ford and General Motors began production of their first-generation electronic carburetors in 1977. They applied these to engines that failed to meet the new emission control standards with carburetors of existing types. The same cars

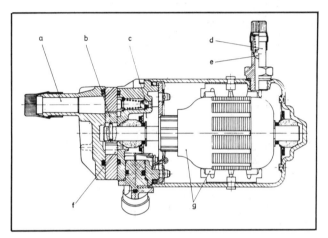

Fuel pump for the D-Jetronic is a roller-cell pump with a cylindrical space in which an eccentrically mounted disc ensures the pumping action and builds pressure in the fuel line. Its rotation produces periodic volume variations at the inlet and discharge ports, while the rollers are forced outward by centrifugal force and form a running seal. The rotor is driven by an electric motor.

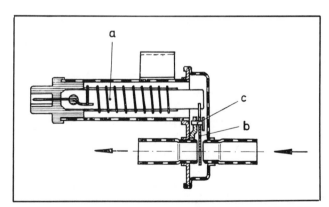

The Bosch Jetronic supplementary air valve works with a bimetallic spring that is electrically heated. It is mounted on the engine at some location where its operating temperature prevails, and it opens to admit supplementary air whenever required. a-BI-metallic spring; b-Revolving disc; c-Center of disc rotation.

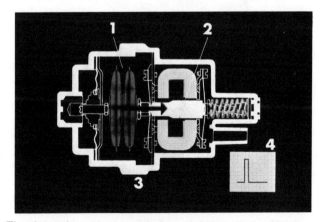

The Jetronic pressure sensor contains an inductive data transmitter connected to an electronic time switch in the control unit. Two evacuated aneroids serve to move the plunger in the magnetic circuit of the transformer and hereby change its inductance. With a closed throttle, the manifold absolute pressure is low, and the aneroids (1) are expanded, moving the plunger (2) out of the magnetic circuit. The inductance is low, giving a short pulse (4).

were also equipped with three-way catalytic converters and closed-loop feedback.

When the EPA (Environmental Protection Agency) again stiffened its standard for 1981, the second-generation electronic carburetors appeared. The GM system was known as CCC (Computer Command Control). Delco made the electronics and Rochester made the carburetor. The carburetor float did not control the fuel flow by conventional means, but sent electric signals to the control unit, which in turn gave orders to a solenoid mounted on the fuel supply line. The control unit compared the float position with input data from manifold vacuum, coolant temperature, throttle plate angle and Lambda-Sond sensors in computing the correct amount of fuel and the spark timing.

Concurrently, Bosch was supplying the microelectronics for a new Pierburg carburetor. First announced in 1981, it has matured over the years into a carburetor marketed as the Ecotronic. The prefix *eco* can be taken as referring to both *economy* and *ecology*.

Volkswagen introduced digital idle stabilization on the air-cooled flat-four engine of the Transporter/ Microbus in 1979. The control unit received input data regarding throttle plate angle and rpm, but it did not handle the fuel-metering task. Instead, the carburetor was set for the leanest possible mixture consistent with conditions. The control then timed the spark (with transistorized ignition and a breakerless distributor) for a steady idle at the most economical level, yet always in readiness to prevent stalling when the torque demand was suddenly increased. Digital idle stabilization also proved itself effective in reducing carbon monoxide emissions.

Use of electronic controls enabled Nissan to develop a chokeless carburetor that was adopted for the 1984 Sentra and Pulsar for delivery in the United States. The Nissan engineers, assisted by the Hitachi carburetor specialists, recognized that electronically controlled carburetors did not allow the same degree of precision in fuel metering on cold-starts and during warm-up as did electronic fuel injection systems. In Nissan's view, this was because mixture enrichment in carburetors was achieved by means of a choke valve. Thus, the elimination of the choke became Nissan's main target.

The choke valve and fast-idle cam were replaced by two solenoid valves and an idle speed control actuator, both under the control of an electronic brain in accordance with input data regarding engine rpm and mass airflow. In addition, the air/fuel ratio and spark timing were calculated on the basis of air temperature, coolant temperature, atmospheric pressure, vehicle speed, transmission gear selection and the air conditioner's on/off status.

The air-fuel ratio was determined by the action of the two solenoid valves. One valve was the same type as that generally used in feedback carburetors and was called the air-fuel valve. It controlled the vacuum in the slow air bleed passage and the fuel supply in the main line on the primary side, according to signals from the Lambda-Sond. The second solenoid was a

Lucas claims to have been the first in the world to announce a digital control fuel management system. The key to the Lucas invention is the large-scale integrated circuit, which is smaller than a postage stamp yet contains over 3,000 components.

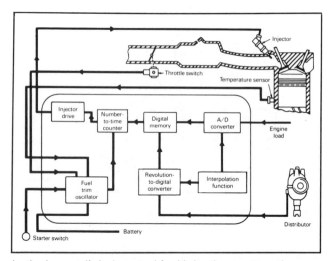

In the Lucas digital control fuel injection system, the control unit gets a reading on engine load from a manifold vacuum sensor. Speed readings are taken from the ignition distributor.

In 1986, Lucas CAV went into production with the LD 100 solenoid fuel injector for multipoint port injection systems.

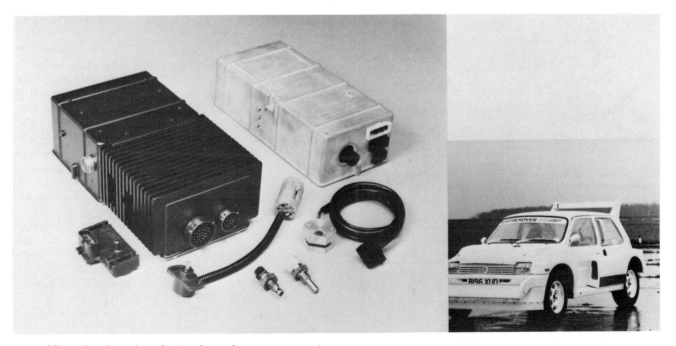

Lucas Micos developed an electronic engine management system for the MG Metro 6R4 V6-4V competition car in 1985.

The Rover 800 appeared in 1986 with its own multipoint electronic injection system, developed by its Fuel Systems branch (formerly SU-Butec).

A throttle body (single-point) injection system was also put in production for the Rover 800, with electronic controls.

fuel enrichment valve with direct control over fuel supply through an open passage downstream of the secondary throttle valve.

The idle speed control actuator consisted of a control diaphragm and a power diaphragm. The power diaphragm was displaced by the combined vacuum from the intake manifold and the leak orifice, and a pushrod attached to its center pushed the throttle valve open. The control diaphragm was subject to vacuum from an electronically controlled vacuum valve in accordance with the engine's target rpm as determined by the control unit. Since there is no vacuum before a cold-start, the throttle plate was automatically set at an adequate angle. Once the engine had started, the actuator began to operate.

Electronic powertrain management and diagnostics

Some expensive cars now combine the automatic transmission control functions with the spark-timing and fuel-metering logic in the same electronic control unit.

BMW could not keep the Motronic for its exclusive use very long. It has spread to other car makers, such as Mercedes-Benz and Opel. And the same basic idea has been picked up by Marelli/Weber in Italy, Renault in France and Lucas-CAV in England.

Despite Renault's early progress in electronic transmission control, it was not until the company founded Renix (in a fifty-one/forty-nine percent partnership with Bendix) in 1978 that the Regie made any commitment to electronic fuel injection. When the Renix system first appeared on the 1984 model R-25 (with the 2.2 liter GTX four-cylinder engine), it was combined not only with electronic ignition but also with the shifting of the optional automatic transmission. Since then, Renault has sold its share in Renix to Bendix-France, which supplies the new R-21 Turbo with an improved electronic fuel-injection-and-spark control system.

The parts-integrated digital fuel injection system of the 1981 Cadillac was the first to offer a true built-in diagnostic routine. It had four programmed types of tests that could be made whenever the situation called for them: engine malfunction tests, switch tests, data display tests and output-cycling tests.

The engine malfunction tests detected system failures or abnormalities. When a malfunction occurred, a warning light went on and a "check engine" read-out appeared on the instrument panel. The control module memory contained a number of trouble codes for each type of malfunction.

If a data sensor failed, the control module inserted a "fail-soft" corrective value in its calculations and continued to operate the engine. If the defect cleared up by itself, the warning light and "check engine" instruction were turned off, but the trouble code remained active in a condition known as "intermittent failure."

The switch tests checked the operation of the switches that provided input to the control module. Data displays were programmed for comparison with an engine that was operating normally. Finally, the output-cycling tests gave a check on the solenoids and lamps.

Since then, BMW has gone a lot further, adding a whole maintenance schedule to its diagnostic functions, which are also more elaborate than Cadillac's. The maintenance schedule indicates when it's time to change oil, get new spark plugs and perform many other routine chores for which the makers usually specify a fixed mileage, regardless of operating conditions. BMW's onboard computer takes all variables into account, so that needless maintenance is postponed when a car does a lot of long-distance travel, and needed jobs are done sooner when a car spends all its time in stop-and-go city traffic, with long idling periods interspersed with brief accelerations.

Recent engine management controls developed in Great Britain and Italy are compatible with such monitoring systems. Lucas-CAV developed a new electronic fuel injection system with sequential delivery to individual solenoid injectors mounted in the ports for the 1985 Jaguar V-12 XJR-6 racing prototype. This system has since also been adopted by Austin-Rover for the MG Metro 6R4 V6 4V rally prototype, on which all control of fuel metering and spark timing is carried out within an Intel 8032 microprocessor.

The electronic control unit directs the fuel supply to six injectors on an individual basis (not grouped). The injectors are standard solenoid injectors supplied with fuel from a constant-pressure rail (80 psi), and each injector takes approximately 2 amps when energized. Opening periods reach twelve milliseconds to supply enough fuel for the engine to develop more than 400 bhp (brake horsepower) at 9000 rpm. A standard Lucas AB14-type ignition amplifier using two Lucas 35C6-type coils and a single high tension distributor is mounted on the front of the engine and driven by one of the four camshafts.

Fuel quantity is determined by the opening duration of the solenoid injector, which is set in a map as a function of engine speed and throttle angle (512 x 8). Lucas chose to use a throttle angle sensor for load input data in preference to a mass airflow sensor or manifold pressure sensor because of the rugged nature of such a sensor and the very fast throttle response that is obtained.

The basic quantities are refined by modifying functions, which take account of ambient air pressure and temperature, engine temperatures and battery supply voltage. These modifiers enable the engine to still produce the best available torque at high altitudes and high ambient air temperatures and also ensure

that starting of the engine under very cold conditions is optimized.

Another particular feature of this system is the ability of the electronics to calculate the amount of fuel required for each cylinder just milliseconds before it is needed, so that each cylinder firing is treated as a new and individual event. This very rapid update eliminates any lag in the system due to out-of-date information and is vital if the full transient performance potential of the engine is to be realized.

Several auxiliary functions are built into the system to take advantage of the power of modern microprocessors. Examples of these are as follows:
• A rev-limit facility to protect the engine from overspeeding.
• Fault detection that monitors each sensor for out-of-limit values. Detection of a suspect sensor leads to substitution by a "now expected" value, and the fault is registered for diagnostics.
• Diagnostic read-out by means of a "check engine" light to indicate the existence of any malfunction and a code identifying the component. The system remembers and recalls the intermittent faults.
• Fuel pump control logic that limits the operation of the fuel pump to a "when needed" condition. In the event of an accident leading to engine stoppage, fuel is shut off to reduce the fire risk.

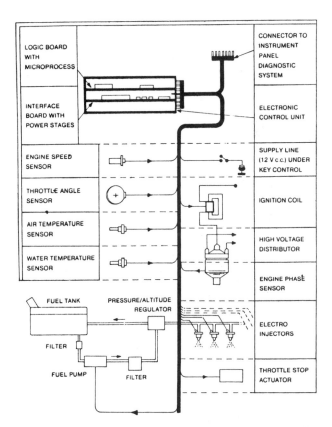

Schematic for the second-generation electronic fuel management system by Spica.

In September 1986, Lucas-CAV announced a new solenoid injector suitable for all aspirated or turbo-charged engines with simultaneous, intermittent or sequential injection. Compared with the usual pintle-type nozzle, the LD-100 features a wider dynamic flow range and improved freedom from fuel deposit plugging. It was developed at the Lucas research establishment at Greenville, South Carolina, where initial manufacturing has been confined.

When Austin-Rover was preparing the 800 (Sterling) series car, two versions of the twin-cam sixteen-valve two-liter four-cylinder engine were planned: one with single-point injection, the other with multi-point injection.

Austin-Rover developed its own single-point injection system from the technology that existed within its SU-Butec division, but it went to Lucas for the multipoint system. There is no interchangeability between the two systems.

In the single-point system, the electronic control unit computes signals from a plurality of sensors to provide fuel injection pulse timing and duration, spark timing and idle speed stabilization.

The system has six sensors:

1. The inlet manifold absolute pressure sensor is located inside the control unit and is connected by a pipe to the inlet manifold.

2. The crankshaft sensor provides engine speed and crank angle signals.

3. The inlet air temperature sensor measures the air temperature at the manifold entry.

4. The ambient air sensor monitors the fresh air temperature to provide data for fuel mixture enrichment for cold-starting.

5. The throttle plate angle potentiometer senses throttle opening and records changes in throttle position.

6. The knock sensor provides signals to the control unit to indicate when detonation occurs and in which cylinder.

Orders from the control unit go to the fuel pump, which feeds low-pressure gasoline to the injectors. An oil pressure switch deactivates the fuel pump in case the engine lacks adequate oil pressure to operate reliably. The injectors provide accurately timed pulses of fuel delivery into the single throat of the throttle body. The throttle body carries a stepper motor to maintain a steady idle speed under varying load conditions, regardless of engine temperature.

A throttle switch indicates to the control unit when to cut off the fuel for deceleration and when to maintain idle speed. A manifold heat sensor switches on the electric manifold heater to promote vaporization during warm-up conditions. A serial diagnostic connector provides a communication link to enable system status to be monitored and diagnosed.

With the single-point system, the engine puts out 120 hp at 5600 rpm, with a peak torque of 162 Nm at 3500 rpm. With multipoint injection, the same engine puts out 140 hp at 6000 rpm, with 179 Nm of torque at 4500 rpm.

The Lucas multipoint system combines port-mounted, solenoid-operated needle valves with the microprocessor-based control unit discussed earlier. It receives input signals regarding airflow, throttle position, vehicle speed, coolant temperature, fuel temper-

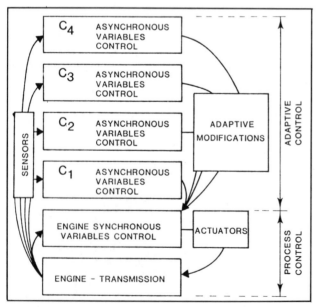

Algorithm of the control logic in the second-generation Spica fuel management system.

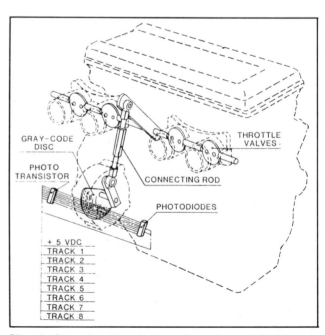

Signals from throttle plate angle and crankshaft angle sensors were coordinated by a complex system of photodiodes and phototransistors.

Control unit for the Volkswagen carburetor with digital idle stabilization.

ature (in the fuel rail), crank angle and rpm. An electric fuel pump delivers pressurized gasoline to the injectors, and fast-idle enrichment is provided by a combined air valve and stepper motor. The throttle potentiometer provides signals to indicate not only throttle position at every moment but also the rate of throttle movement. A serial link connector is used only for diagnosis, allowing the use of specific diagnostic equipment without disconnecting any components. However, the system lacks continuous onboard monitoring in its standard form on the Rover 800 (Sterling).

As a fully owned subsidiary of Alfa Romeo, Spica had the advantage of being able to draw freely on new technology developed by the parent organization. In return, it had an obligation to provide the desired componentry whenever Alfa Romeo scheduled an improved fuel system for production.

About 1975, Spica was instructed to develop an electronically controlled (digital) engine management system for Alfa Romeo car engines. After nearly four years of experimental work, a first-generation system was proposed. Alfa Romeo turned it down following a thorough test program, but took a strong hand in developing a second generation of it.

The four-throat throttle valve angle sensor is covered by an Alfa Romeo patent. It uses a mechanical linkage to a gray-code disc, which is mounted between a photo transistor and a photodiode. The angular change in the throttle plates is faithfully reproduced by photoelectric means and relayed to the microprocessor.

The fuel line pressure valve is also an Alfa Romeo development. It takes full account of atmospheric pressure to maintain the desired air-fuel ratio at any altitude. This valve works with the lowest practicable

line pressure to sustain prolonged injection periods and also to permit the leanest possible air-fuel ratios.

Control of the injector was ensured by a dual circuit with two current levels: a low "switching" level and a high-voltage burst for each valve opening. This setup required highly advanced electronics, especially since the installation space available in Alfa Romeo cars was restricted, and underhood temperatures tended to run high.

When Alfa Romeo was testing control systems for operating its 1500 cc and two-liter four-cylinder engines on two cylinders only under low-load conditions to save fuel, a lot of interesting software was developed in connection with the modular engine concept.

The second-generation Spica system was tested on five different engine types over a two-year period: the four-cylinder inline two liter, the 1500 cc flat four (Alfasud), the two-liter V-6, the US version of the 2.5 liter V-6 and the turbocharged two-liter inline four cylinder. The four-cylinder two liter with Spica fuel injection went into limited production in 1983, followed by the two-liter V-6 in 1984.

The second-generation system included a provision for full vehicle diagnostics, but this never came to application on a production model.

Spica adopted a hot-film airflow sensor and added sensors for clutch engagement and gear selection, with provision for a future automatic start/stop system. Wheel rotation sensors in the driving wheels permitted extension of the control logic to antispin control.

On turbocharged engines, the opening and closing of the bypass valve (waste gate) were placed under command of the same microprocessor. A means of interfacing these with a cruise control system has also worked out.

Since Fiat's takeover of Alfa Romeo, Spica has been given other assignments and its fuel systems activity has been brought to a halt. The group's future needs will be filled by Marelli/Weber.

The components of the Pierburg Ecotronic carburetor.

10

Bosch L-Jetronic

The system known today as L-Jetronic is the direct descendant of the first type of electronic fuel injection systems developed by Bosch in the 1966-68 period, under Bendix patents.

Late in 1969, Citroen adopted the Jetronic for its DS-21, paving the way for Bosch's success in the international market. Saab followed a year later with its 99 E. The Jetronic system was later identified as D-Jetronic. Beginning in 1973, the D-Jetronic was phased out in favor of the L-Jetronic, which took advantage of the tremendous progress made in semiconductor technology over the preceding ten to twelve years.

The L-Jetronic was developed by Heinrich Knapp, who was still chief engineer of fuel injection at Bosch, reporting to Hermann Eisele, director of research and development for diesel and gasoline injection systems. Otto Glöckler was the department manager for the systems, and Dr. Rudolf Sauer was the department manager for gasoline injection components.

As an example of what kind of progress these men had made, the L-Jetronic control unit has only

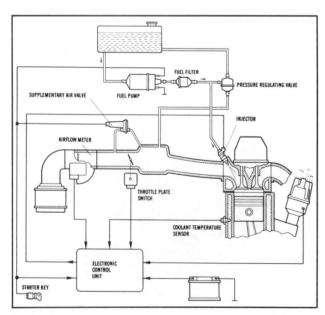

Schematic presentation of the L-Jetronic ignition system.

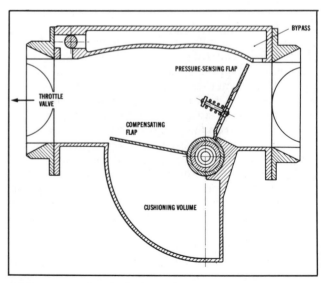

Airflow meter for the L-Jetronic system.

about eighty components, compared with over 300 for the D-Jetronic. This reduction was made possible by such sweeping changes as the replacement of three extensive component groups into simple integrated circuits.

Apart from the three integrated circuits, the L-Jetronic contains only a few semiconductor elements plus a number of condensers and equalizing resistors. The control unit is connected to the main wiring harness by a multipole plug.

This circuit technology offers extreme precision and a high degree of reliability. It also offers wide flexibility in terms of input and output quantities.

Input signals imparting information about the temperature of the incoming air can be introduced, and temperature compensation can be included, by means of a simple resistance network. The system has provision for additional output signals—concerning the control of an exhaust gas recirculation valve, for instance—and Lambda-Sond control (oxygen content feedback from the exhaust system) can be integrated without major complication or expense.

The methods of fuel metering and injection in the L-Jetronic system correspond to those developed for the earlier D-Jetronic. The pressure regulator, coolant temperature sensor and throttle switch are also similar.

In common with D-Jetronic, L-Jetronic provides intermittent injection into the intake ports at low pressure. Unlike D-Jetronic, however, the L-Jetronic system relies on mass airflow metering as the main input variable to determine the amount of fuel to be injected.

In the D-Jetronic system, air metering was accomplished by a manifold pressure sensor, on the principle that while ambient atmospheric pressure prevails in front of the throttle valve, a vacuum is created behind the throttle. This manifold vacuum varies according to throttle position, and it can be used as a parameter to determine the load on the engine as well as the volume of combustion air being admitted into the cylinders.

Bosch subsequently established that manifold pressure is only an approximate measure of the amount of air an engine is consuming.

Airflow metering offers vital advantages for mixture control. It compensates automatically for differences in cylinder filling due to manufacturing tolerances, wear, combustion chamber deposits, alterations in the valve adjustment on an engine or variations in engine speed. It also compensates automatically for differences in exhaust system back pressure (as caused by catalytic converters).

Hermann Scholl served as a group leader on the airflow meter project. The first-generation airflow meter was developed as an independent element. It was an interchangeable assembly mounted upstream of the throttle valve.

Control unit for the Bosch L-Jetronic, with all components mounted on a hybrid-material chassis.

The meter consisted of a box on the air duct, with a pivoted flap standing up in the air passage, yielding to the force of the air pressure. The pivot shaft also carried a potentiometer, which converted the flap angle into an electrical direct current voltage signal.

The relationship between mass airflow and the angle of the flap was chosen in such a manner that the percentage error was kept constant over the full operating range of the engine. The flap was hinged so that the opening area increased when airflow increased. It

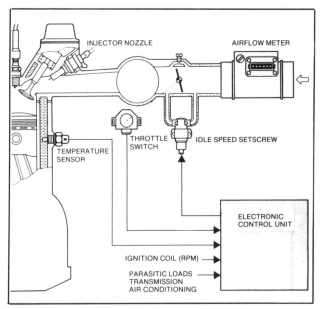

Digital idle stabilization, applied to the L-Jetronic system (with hot-wire airflow meter).

was designed to counteract vortice formation and induced pressure losses throughout its travel.

To avoid overreactions (flagging) in response to sudden changes in the airflow, the flap was spring-loaded against the force of the air pressure. The coil spring had a constant rate, giving a constant balancing force and a constant pressure force.

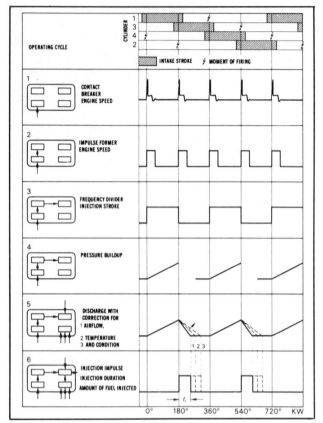

Impulse diagram for the L-Jetronic system.

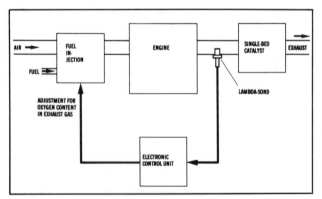

The L-Jetronic system is perfectly adaptable for inclusion of the Lambda-Sond, which measures oxygen content in the exhaust gas. This allows closed-loop engine control, continuously adjusting the air-fuel ratio in accordance with the composition of the combustion products.

The flap was also provided with a damper in the form of a compensating flap, connected by a narrow aperture to a damping chamber. It was mounted on the same pivot shaft as the air-metering flap, and it minimized the effects of fluctuations in manifold pressure on the flap angle. To avoid damage to the airflow meter in case of backfiring in the manifold, the flap had a built-in spring-loaded back pressure valve.

The relationship between the mass airflow and the flap angle was logarithmic, which ensured a nonlinear characteristic in the potentiometer. The logarithmic relationship obtained this way had the advantage of making the airflow meter most sensitive when airflow values were lowest and the highest precision was needed.

The potentiometer offered the additional advantage that the relationship between the flap angle and the voltage could be expressed in a nonlinear pattern, reflecting the actual fuel requirement by choosing the correct geometry for the track.

The track of the potentiometer was arc-shaped with a ratchetlike contour. Its geometry was determined so as to give an inverse relationship between airflow and voltage. Low mass airflow gave high voltage, and high mass airflow gave low voltage.

Because of the exact signal processing in the control unit, the fuel quantity discharged was inversely proportional to the potentiometer voltage. That necessitated the use of a potentiometer with an exponential curve, which caused a steeper curvature of the voltage and made it necessary to split up the track in segments with a parallel, low-resistance series at the junctions between the segments to specify the voltage for the end of each segment. That explains the ratchet-like contour.

The series consisted of a series connection of *cermet* (half ceramic, half metal) resistors of high temperature stability. The voltage relationships at the series divider were so far out of proportion to any temperature or aging effects that might occur at the plastic surface, that the series could be optimized strictly on the basis of maximum durability and minimum noise.

Relatively flat cermet resistors can be trimmed to their final values by a computer-controlled laser beam. These values vary from engine to engine and are stored in the computer. If the car manufacturer wants to change them, they can easily be programmed in from the computer keyboard.

This type of circuit opened the way for combining the notched track with the series divider into one thick-film circuit. All resistance and conductor circuits were combined on a common substrate so that difficult or unreliable soldering points did not exist. This degree of dependability could be obtained only by using the most advanced techniques.

Bosch engineers also took extra pains to test the potentiometer, expecting problems with both noise and wear. But by using a conducting type of plastic with a low friction coefficient in combination with silver-palladium contacts, wear was eliminated as a problem, even at high temperatures.

The voltage from the potentiometer was received in a multivibrator, which also received signals from the impulse former and ignition contact breaker. The signal from the throttle valve and signals from the starting switch and the temperature sensor went into the multiplier stage, where they were compared with the input from the multivibrator. The multiplier's output went into the terminal stage, which directed the injector valves.

The voltage from the airflow meter was a measure of air quantity per unit of time. This voltage determined the frequency for the multivibrator, which was triggered twice per camshaft revolution.

The multivibrator contained a condenser. During the interval between two firing impulses, this condenser was charged at a constant rate so that at the end of the period, the condenser voltage was inversely proportional to the rpm.

The system handled mass airflow variations over a scale of 40:1 and rpm variations over a scale of 10:1. The amount of fuel injected varied only over a scale of 4:1.

The airflow meter in the L-Jetronic had one similarity to the carburetor in that it provided a highly accurate measure of the actual mass of air entering a particular engine under any combination of speed and load. Like a carburetor, it automatically responded to the little differences in combustion chamber volume, port shape and other factors that affect volumetric efficiency.

That sensitivity was good for both fuel economy and emission control. But it had the same weakness that was also inherent with the carburetor: Compensation for changes in air density was not complete, and thus the engine tended to run overrich at high altitude or when the underhood temperature was high.

Theoretical mixture strength is proportional to the square root of air density, which means that all systems relying on airflow metering are subject to altitude effects. Because the air is less dense at higher altitude, the mixture tends to get richer.

For altitudes up to 7,000 feet, this has no practical significance. But US emission control standards specify high-altitude tests. Therefore, the system included an altitude meter, in the form of a barometer converted to a potentiometer. Its signals were fed to the control unit, which then adjusted its data from the airflow meter to compensate for the altitude change.

Since then, Bosch has developed hot-wire (see the section on LH-Jetronic later in this chapter) and hot-film airflow meters, with significant advantages over this initial type.

On the fuel supply side, the L-Jetronic has a number of clever features. The gasoline is delivered from an electric fuel pump, through a filter, to the pressure regulator at a slight overpressure.

The fuel pump is of the roller-cell type, driven by a permanent magnet and a motor. It consists of a cylindrical cavity in which is placed an eccentrically revolving plate. The circumference of the plate has indentations for the metal rollers. These indentations serve as seals (owing to centrifugal force) and create a pumping action, since the volume alternates from maximum to minimum as the eccentric plate revolves.

The pressure regulator is a metal casing divided into two chambers by a diaphragm, which is spring-loaded against the direction of fuel entry by a coil spring. Pretension on the coil spring determines the basic fuel line pressure. The spring chamber has a vacuum line to the manifold. This serves to make the fuel line pressure dependent on manifold pressure, so that the pressure loss should be uniform for all injector valves. When the pressure exceeds the prescribed value, the diaphragm opens an axially adjoining return line, which is not pressurized.

Because the pump constantly delivers more fuel than the engine ever needs, a steady overflow is returned to the tank from the pressure regulator.

Fuel metering is accomplished by electromagnetically operated injectors. Branch lines run from the pressure regulator to the individual injectors. At constant fuel pressure, the amount of fuel injected is proportional to the injector-opening duration. This duration is calculated with a high degree of precision and optimized for every engine condition by the electronic control unit.

The control unit receives a continuous flow of signals from sensors that measure charge air temperature, coolant temperature, engine speed and load.

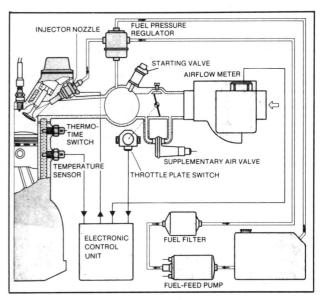

Schematic of the LE-Jetronic system.

The injection pressure is controlled by the overflow pressure regulator, which is balanced on the constant differential between manifold pressure and fuel line pressure. It is generally about 35 psi.

The spring chamber in the pressure regulator is connected with the intake manifold. By this method, it is ascertained that the amount of fuel injected depends only on the duration of injector valve opening.

The injector consists of a body with a jet needle and an electromagnet. The electromagnet is shaped like an armature and placed midway in the body. When the magnet coils receive no current, the jet needle is held against its seat by a coil spring. When the current arrives, the magnet is activated and lifts the jet needle about 0.15 mm (millimeter) from its seat, letting fuel run out through a calibrated annular void. The tip of the jet needle is shaped for spraying the fuel in a pattern for best atomization. Opening and closing times for the needle are about one millisecond.

To miniaturize the electronic switching, the injector valves were arranged in parallel, so that they open and close all at the same time and not in sequence (such as firing order). As a result, the residence time of the fuel injected in each port varies.

To counteract this and try for fully uniform fuel distribution, the injectors are set to deliver the metered amount in two portions: half the amount twice during each revolution of the camshaft.

This solution avoids any fixed relationship between the cam angle and the moment of injection. It also permits triggering of the injection by the contact breaker, so that for every second firing (in a four-cylinder engine), there is one injection impulse.

Some remarks about the injector installation may be important. Fuel deposits on the manifold walls must be avoided in the interest of effective emission control. Therefore, do not spray against the walls, but keep the spray angle within twenty-five or thirty degrees, aimed right at the back of the intake valve head.

The injector valves must have thermal insulation against the manifold to avoid hot-starting problems caused by fuel evaporation in the port areas.

The electronic control unit calculates the fuel quantity on the basis of all input data. The order is given to the fuel injector valves in the form of electrical pulses that actuate them.

These pulses determine the fuel flow per stroke, while the airflow meter indicates mass airflow per unit of time only. The two inputs are coordinated by a division according to rpm. This operation needs an explanation: Trigger impulses are taken from the contact breaker (or a corresponding terminal with breakerless ignition) and led to the multivibrator through an impulse former and frequency splitter stage.

The multivibrator frequency is proportional to the interval between two firing impulses; in other

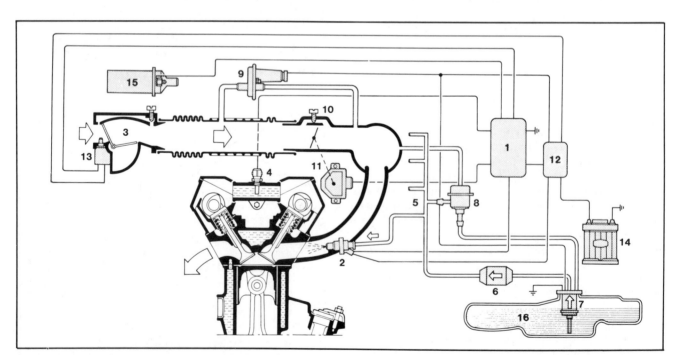

Bosch LE 2 Jetronic installation on the Fiat Croma Turbo i.e. two-liter twin-cam four-cylinder engine, 1986. 1-Electronic injection control box, 2-Injector, 3-Airflow meter, 4-Water temperature sensor, 5-Manifold conveying fuel to the injectors, 6-Fuel filter, 7-Dipped fuel pump, 8-Fuel pressure regulator, 9-Supplementary air valve for cold-starting, 10-Throttle, 11-Accelerator position switch, 12-Remote control breaker, 13-Air Temperature Sensor, 14-Battery, 15-Coil, 16-Fuel tank.

words, it is inversely proportional to the engine speed. Consequently, the mass airflow signal must be divided into the rpm and then converted into a number giving air quantity per stroke.

On their way through the control unit, the pulses from the contact breaker are first transformed from peaks into rectangular pulses. These pulses have sharp edges that switch the charging and discharging current of a capacitor in alternation. The charging current is constant, and the discharging current is determined by the angle of the lever on the airflow meter.

The discharge period emits the next switching pulse. Its duration is dependent only upon the interval between two ignition pulses and the ratio of charging to discharging current. This solution ensures that engine speed is measured in a mathematically correct mode. It also precludes interference caused by temperature fluctuations in the capacitor and voltage variations in the current.

In a second stage, a similar charging and discharging process takes place with these switching pulses. Charging takes place during the pulse, and the charging current is determined by the engine's operating conditions. A resistance wire that senses engine temperature and a throttle switch that signals the load on the engine are connected so as to influence the charging current.

A pulse is also taken at the time of discharge. The discharge pulse immediately follows the charging pulse, reflecting vital data such as load and temperature. The entire pulse goes through an amplification stage, and then its signal goes to the fuel injector valves.

Most of the active components are in operation on an on/off basis, which precludes sensitivity to ambient temperature and the effects of aging.

The operating mode is particularly well suited to the primary function of the control unit, which is to conduct the switching operations triggered by ignition pulses through several converter stages directly to the fuel delivery valves, determining their opening duration and thereby controlling the amount of fuel to be injected.

L-Jetronic automatically copes with transitions, such as when the throttle valve is suddenly opened wide and the engine is faced with the risk of fuel starvation. Because of the immediate reaction to any change in mass airflow, the system needs no additional devices to ensure adequate fuel supply under those conditions.

Mixture enrichment for cold-starts is provided by a magnetic valve located centrally in the manifold.

The starting valve is actuated when the engine temperature is below 15°C, coming into action simultaneously with the engagement of the starter motor. A thermo-time switch turns the starting valve off when fuel enrichment is no longer needed. Because of the starting valve, quick and sure starts can be made in ambient temperatures down to –30°C.

The cold-starting valve is not made for accurate timing but for creating the finest possible fuel spray. The nozzle is of the rotating type and has two tangential bores which promote rotation (by fuel flow). The nozzle gives a fine spray of conical shape, with a spread of no more than forty-five degrees. It has a solenoid valve that is spring-loaded to the closed position. Actuation of the solenoid frees the valve seat, and fuel flow begins. The starting valve is disconnected after about thirty seconds. After that, the control unit provides sufficient enrichment.

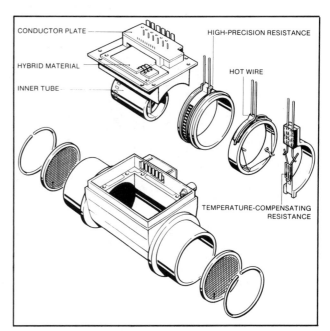

Hot-wire airflow meter.

Bosch hot-wire airflow meter. Cutaway housing reveals cooling element and a total absence of mechanical parts.

Close control of fuel enrichment during warm-up is ensured by a simple temperature sensor. The sensor is mounted in a suitable position in the water jacketing or in the cylinder head and is connected to the electronic control unit.

With very cold charge air, a leaning-out of the mixture occurs owing to the physical conditions in the airflow meter. This can disturb an engine, unless the maker has provided preheating of the intake air. For such cases, a temperature sensor can be placed in the intake duct, to deliver a corrective signal to the control unit.

On cold-starts and during warm-up, the engine needs auxiliary air for idling owing to the higher friction losses that are at play. It thus becomes necessary to ensure a faster idle. Additional air is supplied through a bypass valve, which leads a small amount of air around the throttle valve. The bypass valve's opening area is controlled by the temperature of the engine and the temperature of an electrical heater coil with a bimetallic strip. This additional electrical heating is intended to speed the warm-up and shorten the time when enriched mixture is supplied.

On overrun (when coasting or using the engine as a brake), the engine places specific demands on the air-fuel ratio. These demands are different from the conditions when the engine is pulling.

The L-Jetronic can be set to cut fuel delivery entirely above any chosen rpm and to give a leaner mixture at other speeds. The only condition is that the control unit must be fed information about mass airflow and rpm in the proper context.

To ensure correct engine operation on overrun at very low car speeds, the L-Jetronic opens the bypass valve to admit additional air past the throttle valve and prevent an excessive pressure drop in the manifold.

In many cases, it is preferable to adjust full-load and idle settings separately from the broad spectrum of operating conditions. This is possible with the L-Jetronic, which has a throttle switch that signals the control unit whenever idle or full-load conditions exist, so that the air-fuel ratio can be optimized for those exact conditions. For idle adjustment, the system even includes an adjustable bypass to the airflow meter.

For safety reasons, the L-Jetronic contains an electrical cutoff that prevents fuel delivery when the ignition is turned on but the engine is not running. This is achieved by a switch that cuts the current supply to the fuel pump when the flap in the airflow meter is fully closed. The switch is overriden by engagement of the starter motor.

In an engine whose exhaust is treated in a catalytic converter, it is not possible to limit peak rpm by inserting a centrifugal governor in the ignition distributor, because failure to fire would cause the unburned mixture to overheat the catalyst. In the L-Jetronic an upper rpm limit is ensured by simply giving orders to the control unit to restrict fuel injections to a certain maximum frequency. This gives an effective speed governor without risk for the catalytic converter.

All data regarding the engine functions are collected in the control unit, where they can be made to serve additional purposes. For example, the control unit can be wired to give an audio or visual warning signal to the driver about overspeed, wasteful fuel-flow rates and overheating.

The small dimensions and light weight of the L-Jetronic components make for easy installation in a car. Attention must be paid to provide the airflow meter with rectifier stages on both sides so as to maintain closely defined flow conditions inside it. Otherwise, the engine manufacturer has full freedom to put the elements where most convenient and to optimize the air intake duct for ram effect.

LE-Jetronic and LU-Jetronic

A new-generation L-Jetronic arrived in 1981, the basic system becoming known as LE-Jetronic. By this time, the highest-ranking fuel injection engineering executive at Bosch was a recent arrival, Dr. Hansjörg Manger. Hermann Eisele was his deputy to ensure continuity in the development of gasoline injection equipment. Many other new faces were on the scene, young men schooled in electronics rather than mechanical engineering.

Inevitably, L-Jetronic had to change with the times. For the LE version, new functions were added, yet the control unit was reduced in size. Considerable cost was taken out of the whole system, mainly by simplifying the switching inside the analog control unit and using high-ohm (resistance) injector nozzles.

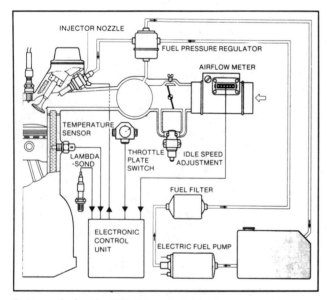

Schematic for the LH-Jetronic system.

The LE-Jetronic was first used on the BMW 525 i and 528 i for 1983.

LE 1 had the usual cold-starting valve with its thermo-time switch. LE 2 had a new electrical cold-start device for mixture enrichment, which was inserted into the intake duct a short way downstream of the throttle plate.

A parallel version, LU—Jetronic, included a Lambda-Sond and all the functions required to meet new and stricter emission control standards in the United States and Japan.

A separate idle stabilization function was added to the LE- and LU-Jetronics in 1983. The idea was to prevent the engine from automatically speeding up in response to sudden increases in load, such as when switching on the air conditioner or activating the power steering system. This function saves fuel, reduces noise and helps reduce the emission level.

L3 Jetronic

Preparations for the third-generation L-Jetronic began in 1977 when I. Gorille was put in charge of digital engine control studies. By using digital systems, Bosch was able to miniaturize the control unit so much that it could be packaged inside the airflow meter assembly.

The LE 3 Jetronic's name was simplified to L3-Jetronic by the time production got under way in 1986. This system was immediately adopted by Citroen for the BX GTi and by Opel for the Omega.

LH-Jetronic

The idea behind the LH-Jetronic was to make fuel metering independent of changes in atmospheric pressure and ambient temperature. This was accomplished by going to a hot-wire type of airflow meter, in which the mass airflow is measured when the air flows past an electrically heated filament inside a tubular housing.

Development of the hot-wire airflow meter began in 1977, and production began late in 1981. Its first application was in the Saab 900 Turbo for 1983.

The filament is a 100 micron platinum wire, which is looped around the passage. This is part of a bridge circuit and is heated to a constant 150°C. A fast rush of fresh air has a cooling effect on the filament, which then demands a higher current to restore its temperature to the prescribed level. The greater the mass of air, the higher the energy needed to maintain a constant temperature in the filament.

The digital microprocessor (in the control unit) measures the supply of electric current to the wire.

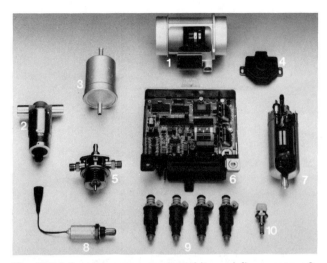

The LH-Jetronic components. 1-Mass airflow meter, 2-Idle speed adjustment, 3-Fuel filter, 4-Throttle plate switch, 5-Fuel pressure regulator, 6-Electronic control unit, 7-Electric fuel pump, 8-Lambda-Sond, 9-Injector nozzle, 10-Temperature sensor.

Since the wire's energy consumption is proportional to the mass of air entering the engine, the microprocessor can accurately quantify the mass airflow. At the same time, the microprocessor receives data on engine speed and temperature, and it can order the exact amount of fuel that best meets the engine's requirements.

The filament is self-cleaning, since it is heated briefly to about 1,000°C every time the ignition is switched off. This high temperature burns off all deposits on the filament so that the filament's sensitivity remains unimpaired.

Shortly after the hot-wire airflow meter went into production, Bosch undertook laboratory experiments with an even more radical concept: a hot film. This is an electrically heated metallic coating integrated with a ceramic substrate, which would replace the filament.

The dimensions of the hot film could be smaller than those of the hot wire, and the number of parts reduced to a minimum. The film strip could be suspended in a small two-piece plastic housing. The film would not need to measure the total airflow, as long as it could be positioned so as to measure a flow mass proportional to the total. Hot film has lower heat loss than hot wire, so the current draw would be reduced.

Mounted in the downstream section of the sensor, the hot film would be immune to deposits or dirt contamination. With fewer components, the hot-film airflow meter would also cost less to produce.

11

Bendix analog system and Cadillac

In 1974, Bendix secured a contract from General Motors to supply electronic fuel injection systems for Cadillac cars to be introduced in the 1975 model year. At that time, Bendix, Bosch and General Motors signed a tripartite supplier-licensor-licensee agreement with the aim of filling GM's fuel injection equipment requirements through the early 1980s.

Because of slippage in Cadillac's program, the introduction was delayed until September 1975, when the Bendix analog fuel injection system was adopted

as standard on the 1976 model Cadillac Seville and as a regular production option for all full-size 1976 Cadillacs.

Bosch supplied some injectors, pressure regulators, throttle position sensors and fuel pumps to Bendix for inclusion in the systems sold to Cadillac. Electronic control units, other sensors, throttle bodies and

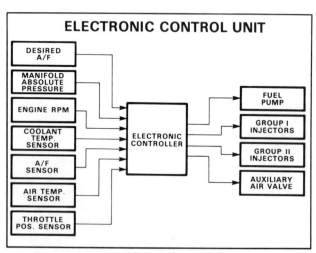

The Bendix concept of the control system evolved into a logical functions chart by 1973. Electronic contouring was used for manifold pressure and engine temperature signals before they entered the pulse-forming network. Afterward, any sensed variable could still be taken into account in the gating of the data.

Bendix electronic control unit from 1974. The experts knew it was only a starting point and were already planning two phases of adapting new technology.

fuel distribution rails were designed and produced by Bendix.

Cadillac had never been tempted to use the Rochester continuous-flow mechanical fuel injection system. The chief engineer, Carlton Rasmussen, was convinced it could only hurt and not enhance Cadillac's reputation. And in the early days of emission controls, Cadillac had no incentive to change from the Quadrajet carburetor.

But after 1970, Cadillac began to suffer from the same drivability problems that affected most American engines at the time, with stumbling and stalling, jerky automatic transmission shifts, cold- and hot-starting difficulties, and running-on (dieseling). Fuel consumption rose sharply in the 1970-73 period, and in March 1973, Cadillac started a serious test program with the latest Bendix electronic fuel injection system.

Bendix had let the Electrojector lie dormant until 1967, when the engineers at Bendix saw an opportunity for its use arising through the pending exhaust emission control standards.

"We started to take a new look at the work we had done on our fuel injection system because we felt that with its accurate fuel control potential, it could contribute to lowering emissions," said John Campbell, then general manager of the Motor Components Division of Bendix.

"After looking at the fantastic developments that had been made during the decade in electronics, we decided to dust off our hardware and data and do additional development work to see exactly what contributions an electronic system could make to lowering emissions."

An engineering group was formed at the Elmira plant in 1968 to study the system's potential. "Our studies indicated that we did have potential for providing auto manufacturers with the tools they needed to help meet emission legislation, while still providing fuel economy, performance and good drivability," explained John Campbell.

In 1970, the group began moving its operations to the Detroit area to keep in closer contact with the automotive manufacturers. In 1972, the Electronic

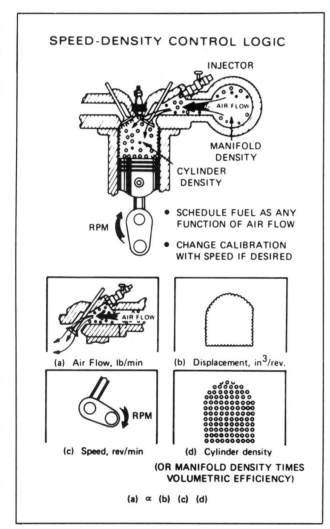

Schematically, the speed-density control logic becomes clear. Fuel metering is based on mass airflow measurement as determined by engine speed and air density. For any given engine configuration, airflow is proportional to the product of cylinder air density and engine speed. Cylinder air density is equal to the product of manifold air density and the volumetric efficiency of the engine (its degree of cylinder filling). On the basis of these factors, fuel flow can be scheduled as any desired function of airflow.

Throttle body for V-8 engine. Its basic function is to adjust the volume of air entering the engine in response to the driver's commands. In addition to the throttle valves which are linked to the accelerator pedal, it includes systems for startup air control and idle-air bypass.

The manifold pressure sensor generates a voltage that is proportional to the level of vacuum in the intake manifold. This unit was only 1.75 inches high, with a square-sided base 1.125 inches long.

Fuel Injection Division was established with headquarters in Troy, Michigan.

By that time, semiconductor technology and electronic systems manufacturing techniques had developed to the point of making the electronic fuel injection system a realistic possibility.

William L. Miron outlined the commercial and technical reasons for the new situation: "With transistors, integrated circuits, thick film, conformal coating and so forth, the cost and complexity has diminished sharply, while the price and complexity of that fourteen-dollar carburetor has risen tremendously. With high-volume production, the cost differential between electronic fuel injection and carburetion is virtually nil, when we consider the entire vehicle system."

The contract with Cadillac became possible, first, because GM's most prestigious car division was under pressure to improve drivability and fuel economy, and, second, because Bendix was not offering just

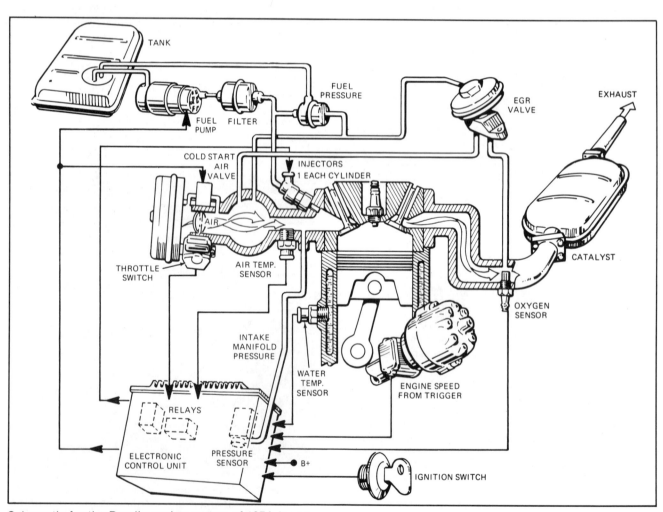

Schematic for the Bendix analog system of 1974. It convinced Cadillac's engineering staff that electronic fuel injection was ripe for production in America.

parts, but a complete electronic fuel management system.

By Bendix's definition, *electronic fuel management* is the use of electronics to sense an engine's fuel requirements as a function of measured engine conditions, to compute the amount of fuel needed to satisfy those requirements and to control the fuel flow accurately in proportion to the air intake for the most desirable combustion level.

Compared with the Electrojector, the new system was different in componentry but not in principle. "Its principles were always sound," a Cadillac engineer asserted. "It was mainly a matter of refining the functions for greater accuracy and reliability, and taking some of the cost out of the system."

Some of the biggest differences were to be seen inside the control unit, in its components and circuitry. It is also noteworthy that fuel line pressure was almost doubled, to 39 psi. With regard to timing, a new adaptation was required because Cadillac had adopted the Delco electronic high-energy ignition system in 1975, and a means had to be devised for enabling the

Fuel-pressure regulator typically operates at a natural frequency of 400 Hz with a gain of 0.043 psi per gallon per hour flow over a flow-rate range of 5 to 40 gallons per hour. In the Bendix system it was preset to maintain a constant pressure of 39 psi above the intake manifold pressure. It bleeds off excess fuel (and pressure) through a flat-plate valve.

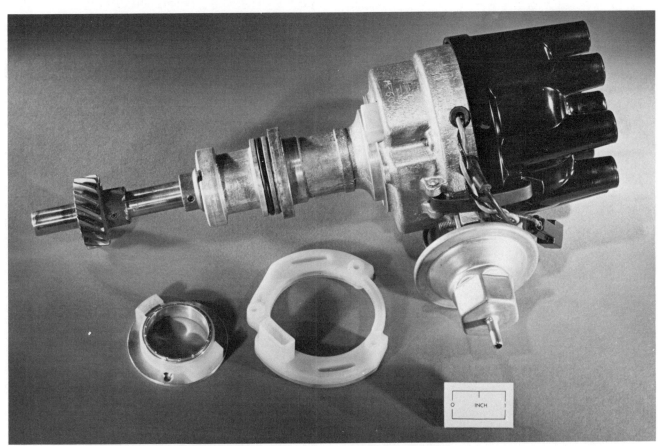

The earlier triggering selector was replaced by a simple speed sensor, feeding a continuous flow of data on engine speed and phasing to the control unit. These data serve to relate the frequency of injector operation and the time at which injection starts. The sensor shown here incorporates magnetic-reed switches which work without any rubbing or sliding contact.

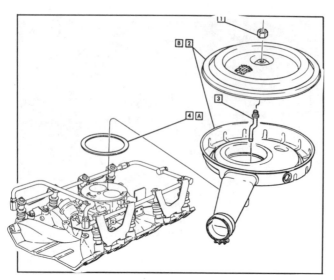

Throttle body and fuel rails on the 1975 Cadillac conformed fully to Bendix designs. The manifold was Cadillac's own.

breakerless distributor to serve as an rpm sensor, with a magnet assembly and a reed switch assembly.

A pair of rotating magnets carried on the distributor shaft, interacting with a pair of reed switches, were encapsulated within the housing cover plate. When the magnets flew past the reed switches, they generated a make-and-break current, accurately counting the engine revolutions. Why two magnets and two switches? Because the injectors were divided into two groups, which were timed separately.

Injection timing was modified so that the system's differences of principle from a continuous-injection system became blurred. The injectors of the four central cylinders were set to spray at the same time, during one revolution; the four cylinders on the front and rear ends received fresh fuel during the following revolution, all at one time.

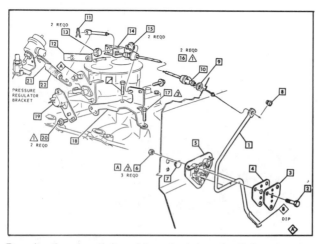

Despite the simplicity of its principles, the linkage to the Cadillac throttle body became a complicated assembly.

The injectors were of the pintle-nozzle type, with solenoid-operated on/off valves. The injectors were inserted in the intake ports and aimed at the back of the valve heads. The amount of fuel varied only according to nozzle-opening duration, which was dictated by the electronic control unit.

The electronic control unit received a continuous flow of data from five sensors: the manifold pressure sensor, the throttle position sensor, the coolant temperature sensor, the ambient air temperature sensor and the engine rpm sensor.

The system was called *analog* because it relied on an analog device for measuring mass airflow. One refinement over the original Electrojector consisted of indexing fuel line pressure to manifold *absolute* pressure.

The manifold absolute pressure sensor had a diaphragm that compared vacuum with barometric pressure. The analog sensor generated a voltage proportional to manifold vacuum, and thereby informed the control unit of variations in speed and load.

This way, airflow rates could be determined within closely defined limits, since airflow is proportional to the product of cylinder air density and engine speed. Cylinder air density differs from manifold air

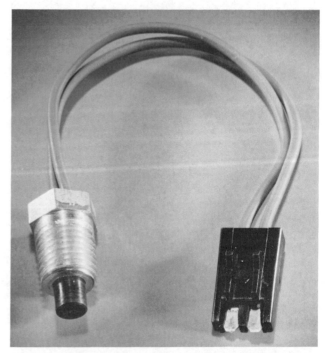

Both the air temperature sensor and the water temperature sensor have the same physical design as shown here. They are two terminal devices comprised of a coil of high-temperature-coefficient nickel wire sealed into an epoxy case and molded into a brass housing. The resistance of the wire varies as a function of temperature, sensor output being linear over the whole temperature range. The voltage drop across it is continuously monitored by the control unit.

density by taking the engine's volumetric efficiency into account.

The throttle body had the task of controlling the main airflow. This was done by a pair of conventional throttle valves linked to the accelerator pedal. The body contained an idle air channel, which had an adjusting screw for setting idle speed (with a warm engine).

A fast-idle valve was installed on top of the throttle body and was arranged to allow supplementary air to enter, bypassing the throttles, during warm-up. The valve had its own thermostat, which gradually closed the valve in accordance with both time and temperature.

Fuel was supplied from an electric booster pump submerged in the tank to the constant-displacement roller-vane-type main fuel pump positioned outside the tank, driven by a direct-current electric motor. The main pump (with its own DC motor) had the task of maintaining a pressure of 39 psi throughout the system, with a pressure regulator referenced to intake manifold absolute pressure.

Cadillac's system was an unqualified success, being regarded as one of the most advanced production car fuel management systems then available anywhere in the world. And the role played by Bendix cannot be overestimated.

An example from the production scheme illustrates how the supplier's involvement overlapped with that of the car company: Bendix received shipments of intake manifolds from Cadillac at its Troy plant; installed the injectors, fuel rail and throttle body; tested and calibrated the complete assembly; and returned the assembly to Cadillac for mounting on an engine.

Meanwhile, Bendix engineers were making rapid progress toward what they thought of as the "ultimate" in fuel management systems.

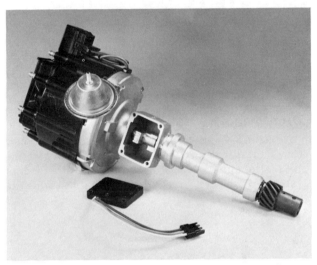

Combined with the Delco high-energy ignition distributor, the speed sensor was mounted on the shaft below the head. It contains a pair of rotating magnets, whose rotational rate is picked up by reed switches in the cover plate and relayed as electronic signals to the control unit.

Fuel injectors for the 1975 Cadillac V-8 are solenoid-operated on-off valves with poppet pintle nozzles. They deliver a narrow-angle spray cone aimed at the intake valve, minimizing manifold wall wetting.

12

Bosch K-Jetronic

The name K-Jetronic is perhaps misleading, because the Jetronic trade name was identified with electronic systems before the introduction of K-Jetronic in 1973. K-Jetronic is not electronically controlled but is instead a mechanical system. In fact, it is one of the simplest existing nonelectronic injection systems. It differs from other mechanical injection systems on three points:

1. Fuel is injected continuously.

2. No mechanical drive is used.

3. Fuel metering is based on mass airflow metering.

Looking at all possible means of eliminating the mechanical drive, Bosch engineers Heinrich Knapp, Reinhard Schwarz and Gerhard Stumpp reasoned that the best way would be to adopt continous injection. This meant developing a method for continuous measurement of the mass airflow as well as continuous metering of the fuel.

Mechanical control systems are not capable of taking account of the air consumption tolerances of the engine, nor can they allow for the effects of partial exhaust gas recirculation (as needed in many emission control systems). Measuring the mass airflow, on the other hand, would—theoretically, at least—enable these factors to influence the fuel metering to the proper extent. The validity of this theory was proved in basic experiments.

Continuous injection was also chosen because a mechanical system is not easily allied with electrical triggering of the injection, which could be done, for instance, by wiring the system to the ignition distributor to get an rpm reading.

The Bosch engineers further reasoned that the essential ingredient for proper control of air-fuel ratios over a wide band of operational conditions was to measure the flow of air drawn into the engine.

This flow cannot be measured by weight, but it can be accurately measured in terms of cubic feet per minute. For proper engine operation, however, the numbers are not relevant. All that is needed is a device that accurately transmits a signal, on a scale arranged in linear progression as the mass airflow increases.

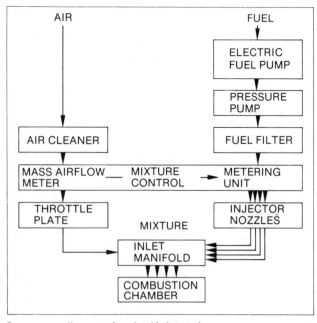

Sequence diagram for the K-Jetronic.

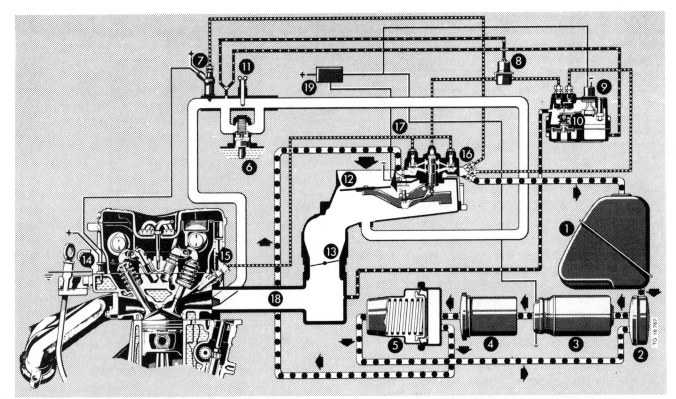

K-Jetronic installation on the 1978 model Mercedes-Benz 280 CE. 1-Fuel tank, 2-Damper reservoir, 3-Electric fuel pump, 4-Fuel filter, 5-Fuel reservoir, 6-Additional air valve, 7-Electric starting valve, 8-Pressure damper, 9-Warm-up control, 10-Bimetallic spring, 11-Idle adjustment screw, 12-Stop valve, 13-Throttle valve, 14-Thermo delay switch, 15-Injection valve, 16-Plunger-type pressure control, 17-Fuel-metering unit, 18-Intake pipe, 19-Relays.

That is basically what a carburetor does—but the K-Jetronic is not a carburetor.

By measuring the airflow mass, the Bosch engineers chose the control principle that promised to give the closest conformity with the actual goings-on in the engine, so as to offer optimum opportunity for controlling exhaust emissions.

Mass airflow is proportional to the cross-sectional area open to it, the air density and the velocity. According to the proven formula, $M = pAV$ when A is the annular area around the plate, V is velocity and p is air density.

With exact measurement of the volume of air flowing into the engine, the fuel metering could be automated, as it is done in constant-vacuum or air valve carburetors. That type of carburetor did indeed serve as a model for the first experimental version, called the KA system (K for continuous, A for driveless).

The air valve or constant-vacuum carburetor works with an airflow range of about 30:1 (i.e., a maximum flow rate thirty times greater than the minimum). By comparison, the venturi carburetor has an airflow range of 900:1, which precludes the degree of accuracy that Bosch regarded as necessary for a competitive injection system.

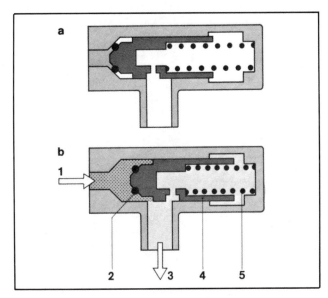

The pressure regulator valve is built into the metering unit. It keeps fuel pressure at a constant 73.5 psi by routing excess fuel back to the tank. Its action is dependent on balancing the fuel pressure against the spring load behind the piston. a-Engine stopped, b-Engine running, 1-Fuel under pressure, 2-Seal, 3-Return line, 4-Piston, 5-Coil spring.

A piston arrangement patterned on that of an SU carburetor was tried in the first stage of experimentation. Many other configurations were also tried and evaluated. But the piston type was less consistent and less immune to friction effects than the lever system that was selected for further development.

The production model airflow meter is built up around a venturi containing a circular metal plate that is suspended on a counterbalanced lever. The suspended plate is forced back by the inflow of air, lifting and tilting from its pivot. The deflection occurs at a linear rate relative to the flow rate, continuously balancing its angle and opening against the force of the airflow.

This linear relationship between mass airflow and fuel metering is the key to maintaining a constant air-fuel ratio regardless of operating conditions in modern engines having a load range of 4:1 and a speed range of about 8:1. After considerable experimentation, Bosch arrived at a device that faithfully translated this relationship into the movement of a single lever.

By making the air passage conical in shape, Bosch ensured that the increase in area is exactly proportional to the change in plate angle. Thus, the relationship between mass airflow and the cross-sectional area will be in direct linear proportion, since airflow velocity remains constant.

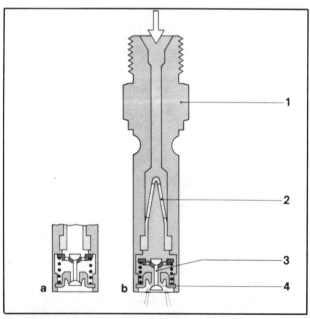

The injector nozzle is installed in a holder that offers a high degree of temperature insulation to prevent vapor lock when a hot engine is switched off. The nozzle is shaped to deliver the fuel in a conical spray formation. a-Closed, b-Open (injection), 1-Holder, 2-Filter, 3-Needle, 4-Valve seat.

The pressure on the plate is determined by the airflow velocity and air density according to this formula: $P = \frac{1}{2}pV^2$ where P is pressure, p is density and V is velocity.

During test work, Bosch engineers arrived at suitable dimensions for the air meter plate:
60 to 80 mm for four-cylinder engines from 40 to 120 hp
80 to 85 mm for six-cylinder engines from 130 to 200 hp
110 mm for eight-cylinder engines up to 300 hp

Increased plate diameter for bigger engines became necessary owing to a loss of accuracy when the plate is forced to extreme angles. That in turn imposes a limit on the pressure drop that can be considered tolerable. Tests have shown that large mass airflow rates and high plate angles cause separations to occur in the airstream, leading to an unsteady lever.

Opposing the pressure of the airflow against the plate by means of a hydraulic constant control force will create a situation in which any increase in the force of the airflow will deflect the plate through an angle that increases the cross-sectional area of the annular passage around it exactly enough to bring the velocity back to its former level.

A bridge piece below the plate works an electric switch that breaks the fuel pump circuit whenever no air enters the duct. Any backfiring through the intake manifold could seriously upset the functioning of the airflow meter. For this reason, the meter has a safety release that allows it to swing over and release the back pressure into the entry duct.

Overcenter travel of the plate is restricted by a rubber cushion, and an adjustable leaf spring sets the plate in its correct neutral position necessary for normal starting.

The airflow meter has a body of pressure-cast aluminum. The plate is also made of aluminum, so that the two are fully compatible in terms of thermal expansion or contraction during temperature variations in the charge air.

The venturi design was carefully laid out to make sure that all of the incoming air participates in bringing pressure on the plate. For the same reason, the usual idle adjustment by air bypass with a screw-type throttle valve was discarded in favor of a mechanical linkage between the air meter and the metering plunger. Under pulsating flow conditions, sensor plate movement is damped by a fixed-caliper pinhole throttle.

The pivot axis for the plate is a stainless steel shaft carried in calibrated teflon-coated bearings. An idle speed roller is mounted on an intermediate arm that is adjustable relative to the main arm. Idle speed adjustment can thus be carried out with the engine running.

The distance between the bearings on the pivot shaft has been stretched to the maximum, with plenty of axial play. The output roller runs on a stainless needle roller bearing with minimal radial runout. The idle adjustment screw is self-locking and carries a polished impact surface. The output lever is a generously dimensioned aluminum pressure-casting.

The proportional movement of this lever from the pressure plate can be transferred to a fuel-metering plunger by mechanical means, with instant reaction to changes in mass airflow.

The control plunger of the fuel-metering unit is under hydraulic pressure from above. The plunger works in a sleeve or barrel provided with a number of very narrow, vertical rectangular slots. It applies a hydraulic force in opposition to the airflow, finding its position at the point of equality between airflow and hydraulic force. When the lever raises the metering plunger, more fuel is admitted into the distributor unit.

The metering and distributor unit is installed upstream of the throttle valve. It should be positioned horizontally, preferably with flexible mounting, so as to avoid acceleration forces exceeding 5 g. Injectors should be installed as close to the intake valves as possible, according to Bosch experts.

A self-imposed condition for the development of the K-Jetronic was that each intake port should have individual shutoff control and should operate at a higher pressure than any pressure head that could normally occur in the manifold or port areas. This condition plays on the requirements for reliable hot-starting, for instance.

The reason for wanting the injector so close to the intake valve is that this arrangement permits satisfactory operation with the leanest mixture and therefore the lowest fuel consumption.

The electrical system is arranged so that the fuel pump begins to function when the starter motor is turned on. When the engine fires and the starter is disengaged, an electric switch on the air-metering lever permits the fuel pump to continue its operation. At the same time, the heating of the resistance wires for the warm-up valve and auxiliary air valve is initiated.

The start-enrichment valve, which is connected to the thermo-time switch in the water jacket, is actuated through the starter motor operation.

K-Jetronic includes two fuel circuits: primary and secondary. The primary circuit is the supply circuit. It begins with a submerged roller-type electric pump, which feeds fuel into an accumulator.

Why an accumulator? Bosch engineers had good reason to put it there. It delays pressure build-up during the starting sequence and after the engine has been switched off. It also helps in damping the noise from the fuel pump. From the accumulator, the fuel flows through a filter into the combined metering and distributor unit. Surplus fuel is routed back to the tank through a constant-pressure regulator.

The accumulator is a tank with two chambers: the spring chamber and the fuel chamber. A single-leaf diaphragm separates the two chambers.

Fuel chamber displacement is about 20 cc, and the spring is arranged to pressurize the fuel contained there during periods of standstill to a level of 37 to 51.5 psi. A spring-loaded valve with a throttle makes it possible to obtain rapid filling of the fuel chamber without needless loss of its effective content when the engine is switched off.

In the metering and distributor unit, both the upper and lower body halves are made from cast iron. The diaphragm is made of steel, which has a similar rate of thermal expansion, so as to ensure freedom from variations due to temperature fluctuations.

The distributor body contains a set of pressure-regulating valves—one for each injector. Each valve

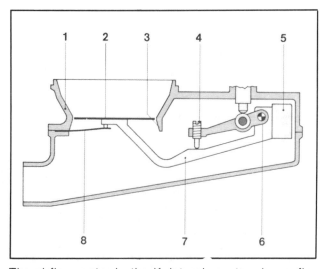

The airflow meter in the K-Jetronic system has a flap mounted on a pivoted and counterweighted lever. Lever motion, modified by the mixture control screw, is transmitted to the metering unit by a plunger-type follower. a-Flap

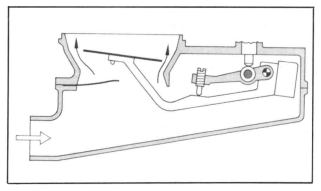

closed, b-Flap open, 1-Body, 2-Flap, 3-No-load area, 4-Mixture control screw, 5-Counterweight, 6-Pivot point, 7-Lever, 8-Leaf spring.

consists of two chambers. A steel diaphragm separates the chambers as well as the primary and secondary circuits.

The secondary circuit is the distribution circuit. It starts with the distributor unit and ends with the injectors in the ports. Fuel flows through a ring-channel in the distributor unit, with ports to the lower chambers of the pressure-regulating valves.

The sleeve and pressure regulator valves are made of stainless steel. The plunger has undergone electroerosive treatment. Inside this unit, there are no adjustment possibilities. Everything must meet specified tolerances in production.

Because of the difference in forces acting on the air meter in four-, six- and eight-cylinder engines even with identical system pressure, Bosch engineers found it necessary to adopt valve pistons of different sizes:

10 mm for four-cylinder engines
13 mm for six-cylinder engines
17 mm for eight-cylinder engines

By using large cross-passages between the individual chambers, the valves operate with exact parallel switching without pressure loss. The entry chambers, following the plunger sleeve, must have sharply defined separations, so as to avoid mutual spill-over. That has been achieved with radial-sealing O-rings for each slot.

To protect the pressure-regulating valves, a nylon filter with a mesh width of twenty-five micromillimeters is installed before the distribution slots.

Pressure in the ring-channel is maintained at a constant 69 psi by a regulator included in the primary circuit. Short, tubular valve seats are mounted in the upper chambers of the pressure-regulating valves of the secondary circuit. Their lower ends bear against the diaphragm, which responds to variations in pressure differentials between the upper and lower chambers. The deflections of the diaphragm dictate the cross-sectional open area of the valve.

By deflections of the diaphragm, the pressure differential between the two chambers is maintained at 1.5 psi, a very low figure which permits extremely small port slots, only 0.1 to 0.2 mm in width.

Fuel enters the valve bodies in quantities determined by the metering plunger. The fuel is then admitted through the slots in the metering sleeve to the discharge ports of the pressure-regulating valves.

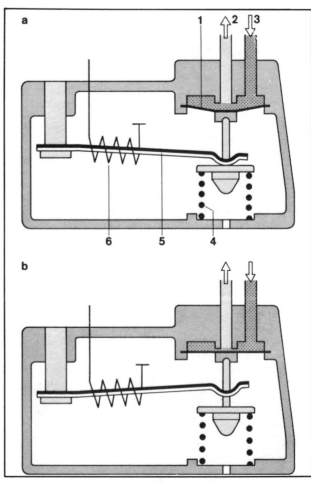

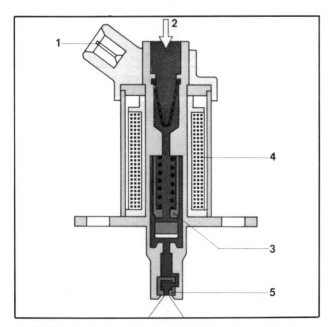

Cold-starting valve in the K-Jetronic system is electrically actuated from terminal (1). Fuel enters at (2) and is filtered on its way to the nozzle (5). Magnetic windings (4) pull on the armature (3) to unseat the valve when cold-start enrichment is wanted.

The warm-up valve cuts back on enrichment gradually as the engine builds up to normal operating temperature. With a cold engine (a), the bimetallic spring (5) is electrically heated (6) from the moment the engine is started. It bends down to let the diaphragm (1) yield to control pressure (3) and open a bigger volume opposite the return line (2). At normal temperatures, the spring is raised, and the diaphragm is in neutral position.

Pressure in the secondary circuit varies between 7.5 and 51.5 psi. A pressure of 48.5 psi is required to open the valve in the injectors. The injector valves include a miniature nylon fuel filter and a chatter valve that senses low flow rates and starts vibrating to ensure adequate atomization of the fuel under such conditions.

The injector nozzle is basically the same as in older injection systems with inline pumps. It is a forward-opening spray valve with a ball-type seat and spherical suspension from a spring-loaded base. Each injector is protected by a nylon filter with a mesh width of fifteen micromillimeters.

Miniaturizing all moving parts has raised their natural frequencies high enough to permit minute quantities of fuel to be injected. At idle, a quantity of 5 cc per minute enters the needle of a reed valve whose natural frequency is 1,500 cycles per second.

Stable and frictionless self-centering of the needle is ensured by its short construction and interface with the spring, which is shaped for it. The valves are thermally insulated by O-rings in plastic carriers. Maximum valve temperature does not occur until the engine is switched off, and it should not exceed 95°C.

The intake manifold should be laid out in accordance with three requirements:
1. Optimum cylinder filling
2. Uniform mass airflow to each cylinder

3. Adequate length for smooth transitions

Point 3 means that for easy transitions from one set of operating conditions to another, the volume of the manifold section between the intake valves and the throttle valve should be equal to or 1.5 times the engine displacement.

Downstream of the throttle valve, the manifold contains a fuel spray valve designed to provide mixture enrichment as necessary for cold-starting. It is electrically operated by an automatic switch that registers both time and temperature signals.

The cold-start enrichment valve is connected to the primary fuel circuit by open lines. A third circuit has the vital role of providing an input on the metering plunger in accordance with the temperature rise during engine warm-up. Pressure in the warm-up circuit is controlled by a pinhole throttle connected to the control circuit. This throttle valve is opened and closed by the cold-start enrichment valve.

Fuel flows to the warm-up valve through an open line from the cold-start enrichment valve, and a bimetallic spring keeps the warm-up valve open during cold-starts.

The bimetallic spring is heated by an electric resistance wire. This heating makes the spring return to its inactive position within a predetermined time (regardless of coolant temperature) and close the warm-up valve.

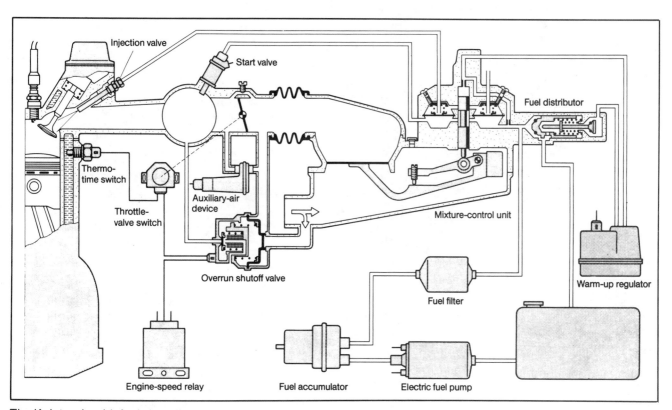

The K-Jetronic with fuel shutoff on the overrun permitted a fuel savings as well as cleaner exhaust.

During warm-up, the valve relieves hydraulic pressure to the metering plunger. When normal operating temperature is reached, the valve closes to maintain a constant hydraulic load on the plunger.

Owing to its basic principles of operation, K-Jetronic had to operate without air containment, as used with electronic fuel injection systems. This meant that different mixture preparation methods had to be developed.

Air containment ensures excellent mixture preparation at idle and in the lower part-load regions. However, it gives no assistance to mixture preparation with increasing absolute pressures, such as in full-load or transient conditions.

Variations in equivalence ratios occur during transitions. For instance, the charge becomes rich after sudden opening of the throttle; that is necessary and corresponds to the action of the acceleration pump in a carburetor. After closing of the throttle but no drop in rpm, the charge becomes leaner. With a sudden drop in rpm on a closed throttle, the charge becomes richer. These changes are mainly due to the conditions existing in the intake manifold between the throttle valve and the intake valves.

The airflow meter accurately registers the change in filling that accompanies a change in load (manifold vacuum). For instance, when the throttle is suddenly opened, the airflow meter must admit the air quantity demanded by the cylinders plus the air quantity needed to raise the manifold pressure to its new level.

The most difficult condition for the system to cope with is a sudden opening of the throttle shortly after a cold-start, from low-rpm operation. Under this situation, a large part of the fuel injected will end up as raw fuel, gravitating to the gutters of the manifold. The mixture will be extremely lean. The engine will not accelerate, and flameout may occur. The engine may stall.

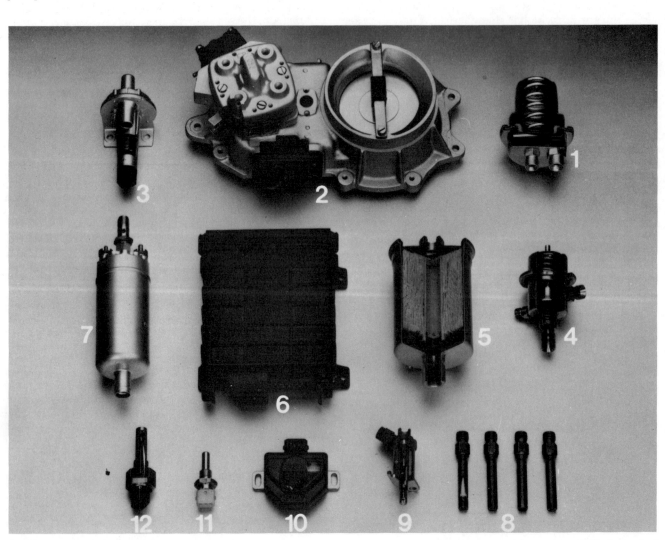

The components of the KE-Jetronic system (minus fuel lines and wiring). 1-Fuel accumulator, 2-Mixture-control unit, 3-Auxiliary air regulator, 4-Fuel pressure regulator, 5-Fuel filter, 6-Electronic control unit, 7-Electric fuel pump, 8-Injection valve, 9-Startvalve, 10-Throttle-valve switch, 11-Temperature sensor, 12-Thermo-time switch.

In a series of experiments, Bosch has demonstrated that the average deviation from the chosen fuel equivalence ratio can be reduced by two methods. The first is to increase airflow velocity across the airflow meter. The second is to include a damping zone between the airflow meter and the throttle valve. Both methods serve to avoid airflow separation through the airflow meter.

The K-Jetronic may have its minor drawbacks, but the system is not running any immediate risk of large-scale displacement by L-Jetronic or newer systems. It is extremely versatile, giving satisfactory results on engines of widely different types, and is fully compatible with turbocharging, for instance. Above all, it has a tremendous cost advantage over all other fuel injection systems, and for markets with strict antipollution laws, it can compete on equal terms with carburetor systems.

Porsche chose K-Jetronic to replace Bosch mechanical (intermittent) fuel injection on its 911s in 1973. In 1975, Mercedes-Benz chose K-Jetronic as a replacement for the D-Jetronic—a backward step in technology, on the surface, taken for the compelling reasons of cost, reliability and ease of servicing. Saab was using D-Jetronic on its US-version 99 for emission control reasons, but adopted K-Jetronic for the new 99 Turbo in 1977.

Beginning in January 1983, K-Jetronic became available with fuel shutoff on the overrun on the Audi 80, 100 and 200; BMW 318 i, 323 i and 520 i; and Volkswagen Golf GTI, Scirocco GTI, Jetta GLI and Passat GLI.

Bosch claimed fuel savings up to five percent for this innovation, which was also offered as a retrofit kit. The main components were an electric switch on the throttle valve shaft, an rpm relay and a one-way valve. When the driver's foot was removed from the accelerator pedal, the switch would complete a circuit that closed the fuel valve completely—as long as the engine was running at a speed of at least 1200 rpm. To prevent too-frequent on/off switching, the valve was actuated with a hysteresis delay of about 300 rpm. When coolant temperature was below 35°C, the fuel supply was uninterrupted, to gain faster warm-up.

KE-Jetronic

During 1981, Bosch began to cultivate the idea that the hydromechanical K-Jetronic could be sub-stantially improved by fitting an electronic control system on it. Responsibility for the development was given to Wolfgang Maisch.

A production version was ready in the spring of 1983. But instead of being a low- to medium-cost installation, the KE-Jetronic became Bosch's most expensive injection system, priced above the L3- and LH-Jetronics.

The first standard application of the KE-Jetronic was the Cosworth-developed engine for the Mercedes-Benz 190 E 2.3-16, introduced in January 1984. Next, Rolls-Royce adopted KE-Jetronic for the 6.75 liter V-8 of the Silver Spur and Bentley Mulsanne Turbo R in September 1986.

Going to electronic control offered certain opportunities for simplification. The KE-Jetronic digital control unit received input data concerning engine rpm (from the ignition distributor), mass airflow, coolant temperature, throttle plate angle (load), starter switch, altitude compensator and knock sensor, which meant that all mechanical measuring devices could be eliminated.

The system's only output came in the form of instructions to a small pressure valve attached to the metering unit. This pressure valve made redundant the hot-start reservoir of the K-Jetronic, though the electric cold-start valve upstream of the injector nozzle was retained.

The pressure valve regulates the pressure (between 2.5 and 25 psi) against the delivery slits in the metering piston. That in effect determines the amount of fuel. Injection remains continuous.

New injectors with thermal air insulation (to prevent fuel evaporation inside the injector) were developed for KE-Jetronic. Leak-air from the intake manifold is drawn off and ducted to the cylindrical jacket around the injector core, where it circulates before premixing with the fuel. This method is also claimed to reduce the size of the fuel particles from 0.04 mm to 0.01 mm.

In contrast with K-Jetronic, the KE-Jetronic is eminently suitable for combination with Lambda-Sond (closed-loop feedback) and digital idle stabilization. Fuel cutoff on the overrun is standard. In case of electronic failure, the system has drive-home capacity in that only the fine-tuning is lost.

13

Bosch Mono-Jetronic

About 1978, Volkswagen research engineers began to investigate what the performance loss might be if the four injector nozzles of an L-Jetronic installation on the 1.6 liter Golf engine were removed from the ports and relocated at the manifold inlet. Perhaps just one nozzle could do the job?

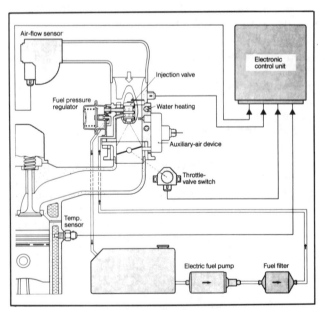

In search of the simplest possible fuel injection system (and the lowest cost), Bosch developed the Mono-Jetronic. The individual injectors in each intake port have been replaced by a single injector mounted at the entrance to the throttle body. The airflow measurement device is of the same type as in the K-Jetronic system.

The incentive came from the constant pressure to cut cost. Compared with a conventional carburetor, even single-point injection with electronic control could be counted on to give a more homogeneous air-fuel mixture and thereby make the engine more economical without loss of drivability.

The engine specialists at Wolfsburg went to work. The first series of tests indicated that when the four injector nozzles were moved far enough back from the ports, they had to be positioned very close together. It became obvious that they could, in fact, be replaced by one single injector.

Bosch had no doubt had the same idea earlier, and had undertaken the basic development of a single-point injection system. But from Bosch's viewpoint, this system was a threat to its price-conscious K-Jetronic market. Bosch made every effort to make sure its customers saw the single-point injection system strictly as an alternative to the carburetor: one step up from the carburetor, several steps down from the K-Jetronic or L-Jetronic.

When Volkswagen's inquiry arrived at Bosch headquarters, it was easy to give quick service. The system already existed, and the necessary hardware was simple to produce.

The system that was later put on the market as the Mono-Jetronic was built up around a mixing unit with an electromagnetically operated injector valve positioned immediately above the throttle plate. The mixing unit body is equipped with a bypass for auxiliary air and also carries a fuel pressure regulator. The fuel-metering calculations are performed electronically on the basis of load (throttle plate angle) and engine speed, mass airflow and coolant temperature.

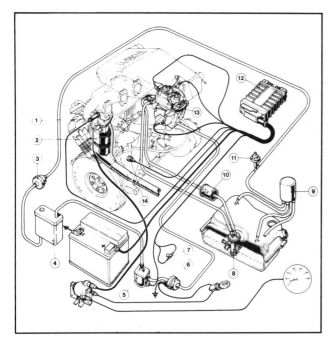

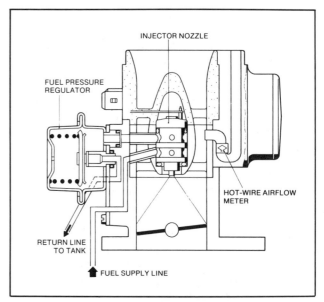

Bosch Mono-Jetronic with hot-wire airflow meter.

Mono-Jetronic installation on the Fiat FIRE 1000 engine for the Uno. 1-Lambda-Sond, 2-Three-way catalytic converter, 3-Fuel tank vent, 4-Fuel vapor filter, 5-Mixing valve, 6-Blocking valve for idling, 7-Resistance, 8-Electrical fuel pump, 9-Fuel vapor separator, 10-Fuel filter, 11-Fuel-vapor-flow control valve, 12-Electronic control unit, 13-Throttle body, 14-Coolant temperature sensor.

Of course, with single-point injection there is no need for timing of the fuel delivery. Since the L-Jetronic system works under constant pressure and only the opening duration of the injector valve can produce variations in the amount of fuel delivered, the

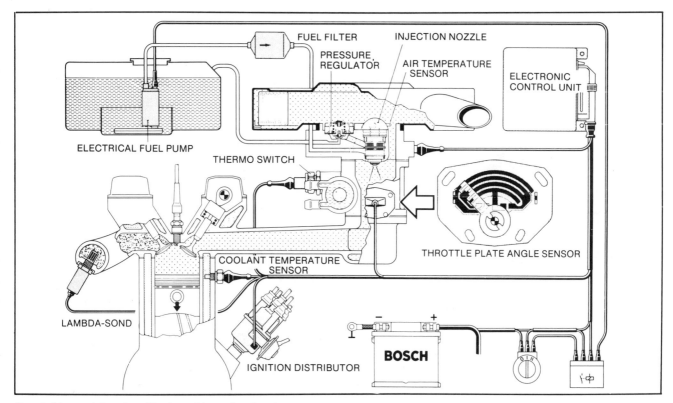

Bosch Mono-Jetronic with closed-loop feedback.

principle of intermittent injection was adopted for the Mono-Jetronic.

Instead of the usual injector nozzle, Bosch developed a fast-switching low-pressure valve. This is mounted directly above the throttle plate at the point of maximum airflow velocity. The valve is operated by a solenoid and cooled by the fuel. The pressure is automatically regulated within a narrow band of 10 to 14.7 psi, which has permitted the use of a lower-cost pump and filter.

Early Mono-Jetronic prototypes used the flap-type airflow meter, but the production versions have been equipped with the hot-wire sensor, which is simpler, smaller, lighter, less expensive and easier to install.

Mono-Jetronic production got under way in 1984, with Volkswagen as the first user. It went into the 1985 Polo with the lead-free-gasoline version of the 1.3 liter engine, which featured a catalytic converter and Lambda-Sond. The standard 9.5:1 compression ratio was retained, and maximum output was undiminished at 55 hp.

Ford did the same for its 1986 lead-free-gasoline Fiesta, which also comes with a catalytic converter and Lambda-Sond. But Ford found it necessary to drop the compression ratio of the 1.4 liter CVH (Compound Valve-angle Hemispherical) engine from 9.5 to 8.5:1, losing 5 hp in the process (70 hp instead of 75 hp).

For the Polo G 40 coupe introduced in 1985, Volkswagen combined spark timing with the Mono-Jetronic control unit, which gave birth to a new system: Digifant. Equipped with the G-lader (spiral compressor), this version of the 1.3 liter engine puts out 112 hp for lead-free gasoline or 115 hp for premium-grade gasoline.

In December 1987, Fiat made Mono-Jetronic available on the FIRE (Fully Integrated Robotized Engine) 1000 for the Panda, and Uno. This lead-free-gasoline unit has reduced compression (9.0 instead of 9.8:1), catalytic converter and Lambda-Sond, but undiminished output (45 hp from 999 cc), though maximum torque has fallen from 80 to 74 Nm, with a 500 rpm (penalty) rise in peak-torque speed.

14

Bosch Motronic

First released by BMW for the 732 i model in November 1977, the Motronic system was not a new type of fuel injection, but a combination of electronic spark timing with the latest digital version of the L-Jetronic.

The roots of electronic ignition go back further than you may think. Delco-Remy began experiments with capacitive discharge in 1936-37, tests with in-car systems began in 1948 and by 1955 Delco-Remy was using semiconductors.

But the first real-life application was made by Lucas. After four years of development, a transistorized ignition system with a breakerless inductive-pulse pickup and a transformer-type spark generator was fitted on the BRM and Coventry Climax Formula One racing V-8 engines in 1962.

Transistor-assisted ignition systems were available on Pontiacs in 1963 and on Fords in 1965, but none were easily adaptable to centralized electronic control. Such systems were not offered until 1971-72, when Chrysler took the lead—but that was long after electronic fuel injection had come on the market.

Bosch built up its own rich background in electronic ignition. The company began supplying a solid-state breakerless digital system to Volvo in 1975 and to Mercedes-Benz in 1976.

At BMW's request, Bosch then undertook the task of putting the ignition functions into the fuel injection control unit. That led to the development of the original Motronic.

In previous electronic applications, the ignition system was adjusted to give the best-timed spark for sure firing under all conditions. However, this could be accomplished only within the framework of the

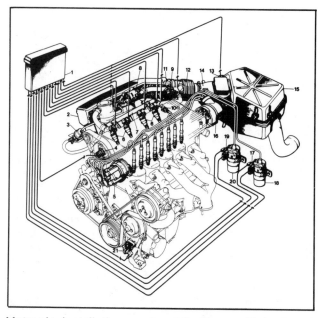

Motronic installation on the Alfa Romeo Twin Spark engine, an inline four-cylinder unit with two plugs per cylinder and variable inlet valve timing. 1-Motronic ignition-injection control unit, 2-Intake capacity, 3-Fuel pressure regulation, 4-Variable inlet valve timer, 5-Timing side spark plug ignition distributor, 6-Electric injection, 7-Fuel distributor, 8-Constant-idling actuator, 9-Minimum-maximum throttle opening, 10-Engine coolant temperature sensor, 11-Throttle unit, 12-Intake duct, 13-Intake air temperature sensor, 14-Intake airflow meter, 15-Air filter, 16-Flywheel side spark plug ignition distributor, 17-Coil power module, 18-Flywheel side spark plug ignition coil, 19-Coil power module, 20-Timing side spark plugs plus ignition coil, 21-Revs and drive shaft phase sensor.

Components of the Motronic system. Grouped around the control unit, clockwise, are the fuel filter, ignition coil, distributor cap, six injectors, fuel pump, temperature sensor and thermo-time switch, pressure-regulating valve and cold-starting valve, throttle valve switch, supplementary air valve and airflow measurement unit.

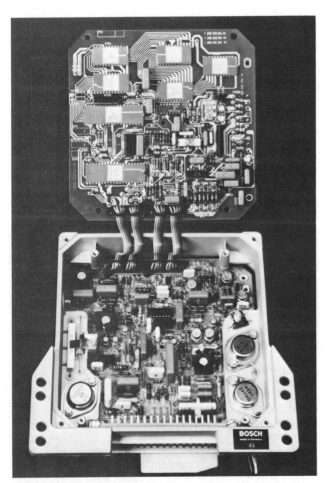

Enlarged view of the electronic control unit for the Motronic system.

adjustment angles controlled by the centrifugal and vacuum-operated spark-advance mechanisms on the ignition distributor. Such applications were not able to take account of mixture strength; consequently, the fuel-metering system and the ignition system occasionally found themselves working at cross-purposes.

The Motronic unites signals of all engine functions and integrates the control of spark timing and fuel metering in the same microprocessor. This goes a long way toward keeping the engine optimally adjusted to all possible speed and load combinations. It also allows the engine to come under optimal and continuous adjustment with regard to efficient combustion (minimal fuel consumption), the composition of the exhaust gases, the car's performance potential and the car's drivability.

Following the switch from analog to digital control, it was possible to retain all the sensors from the standard L-Jetronic. One new sensor was added, the combined rpm counter and timing mark sensor.

This is an induction-type sensor with a permanent magnet. It carries an annular gear whose teeth modulate the signals from the emitter so as to set up an approximately sinusoidal-wave induced current. This current gives an engine speed reading, which is needed for both fuel metering and spark timing. A pickup on the annular gear (or a separate disc rotating at the same speed) registers with the fixed timing marks in sequence and produces another induced current with intermittent signals.

The airflow meter itself is the same as in standard L-Jetronic systems. But for the Motronic, the potentiometer voltage does not begin a declining hyperbolic curve as the plate angle increases. Instead, the voltage undergoes a linear rise in the same situation. This change was made necessary by the adoption of a digital microprocessor.

The electronic control unit is composed of a central processing unit that handles the logic and calculations, two memory banks and an input/output cell that handles all communications with the outside world.

One memory is a read-only memory (ROM) and the other a random-access memory (RAM). When the programs stored in the memory banks come into play, engaging the logic elements, their output is fed back to the switching circuit. From there it emerges in the form of physical impulses for fuel metering, injection timing and spark timing. As they emerge, the signals are weak, but they pass through an amplifier that enables them to actuate the ignition coil and the injector valves.

As used by BMW, Motronic is programmed to cut off fuel delivery on a closed throttle. Only when the engine speed has been brought down to 1200 rpm does the control unit instruct the injector valves to resume fuel delivery. This setup is calculated to save between three and five percent gasoline without affecting the car's behavior or the driving tasks at all.

Motronic is totally maintenance-free. Tune-ups are eliminated, and all the car owner has to remember is to change spark plugs at prescribed intervals. BMW has even developed a system that enables the same microprocessor to calculate plug life on the basis of actual operating conditions plus a time factor, instead of simply counting mileage. The computer then gives a reminder in the form of a read-out on the instrument panel.

Over the years, Motronic has evolved by keeping pace with the evolution in Bosch fuel injection technology. It now exists in versions available in L3- and LH-Jetronic specifications.

Since April 1987, Bosch has also offered a KE-Motronic, based on the KE-Jetronic system, with knock sensor and Lambda-Sond. KE-Motronic can

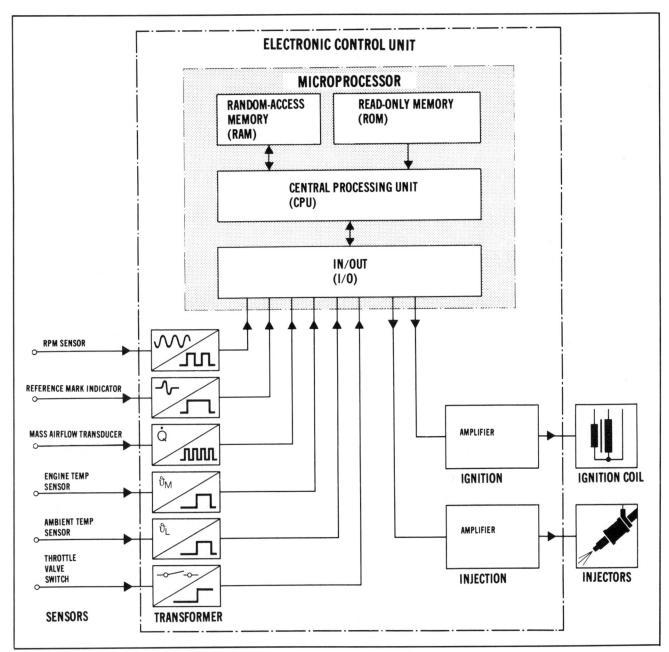

Switching diagram for the Motronic system. The central processing unit is coupled to two memories, RAM and ROM, and an input/output element. Inputs come from (top to bottom) the rpm sensor, reference mark indicator, mass airflow transducer, engine temperature sensor, ambient temperature sensor and throttle valve switch. Outputs go to injectors and the ignition coil.

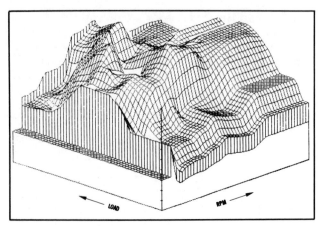

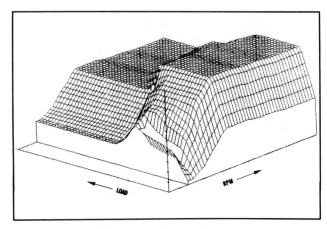

Spark-advance map for the Motronic-equipped engine shows far greater sensitivity to load, up and down the full speed range, and therefore contributes to more efficient combustion (more power for less fuel).

Spark timing with electronic ignition varies in a simple three-dimensional pattern, according to load and rpm.

also be expanded to include digital idle stabilization, electronic fuel tank ventilation and an onboard diagnostic system, according to the client's (the car maker's) specifications.

Bosch's "drive-by-wire" electronic gas pedal was also perfected in 1987 and became standard on the BMW 750 i. The mechanical throttle linkage is replaced by electric wires direct from the accelerator to the Motronic control unit. This unit is programmed with a fail-safe logic, which in case of any malfunction ensures the return of all functions to idle settings.

Together with Bosch and ZF (Zahnradfabrik Friedrichshafen), BMW has also perfected a four-speed automatic transmission whose control functions are handled by the Motronic microprocessor. This is not unique.

Renault was the first out with an electronically controlled automatic transmission, which was offered on the R-16 in 1969. Toyota had its Electronic Automatic Transmission ready in 1970 and combined it with Bosch-licensed Nippondenso electronic fuel injection in 1971.

The BMW-Bosch-ZF system is indubitably the most advanced at the moment. It provides the four-speed automatic transmission with three separate shift programs.

These programs are controlled electronically in full coordination with fuel metering and spark timing and in accordance with manual settings for E (Economy), N (Normal) and S (Sport). For a given gas pedal position, E gives the earliest upshifts and the most leisurely response to kickdown. S lets the engine rev higher before shifting up, makes early downshifts and offers lightning response to kickdown. N is a compromise, about halfway between E and S.

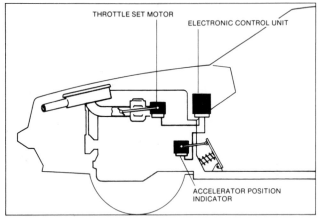

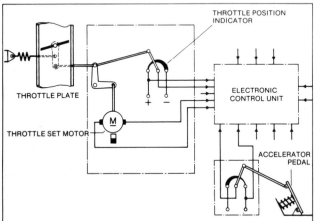

Electronic accelerator pedal. Motronic is available in combination with "drive-by-wire" (as on the BMW 750 i).

15

Bendix: The French connection

The French connection was Renix, a contraction of the names of two partners in a joint venture: Renault and Bendix.

Renault was not a newcomer to the electronics business; in 1966, it had developed the electronically shifted automatic transmission, which became optional on the 1969 model R-16. In 1972, the company started a big electronics program to study the broad field of potential applications in automobiles. Out of this program came an alliance with Bendix for access to its technology and production know-how. This association resulted in the formation of Renix in 1978, with a small factory in the Toulouse area of France.

In 1980, the Renix factory was producing the control unit for the automatic transmission, the regulator for the Normalur cruise control and the integral electronics for Renault's ignition system. The following year, a fuel injection system was added. It was based on some recent work done by Bendix in a renewed effort to reduce the cost of electronic fuel

Fenix-equipped Renault 2.2 liter four-cylinder engine (multipoint version) for the R-25 GTX. With a compression ratio of 9.9:1, it put out 123 hp at 5250 rpm.

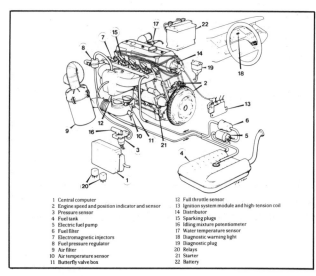

1	Central computer	12	Full throttle sensor
2	Engine speed and position indicator and sensor	13	Ignition system module and high-tension coil
3	Pressure sensor	14	Distributor
4	Fuel tank	15	Sparking plugs
5	Electric fuel pump	16	Idling mixture potentiometer
6	Fuel filter	17	Water temperature sensor
7	Electromagnetic injectors	18	Diagnostic warning light
8	Fuel pressure regulator	19	Diagnostic plug
9	Air filter	20	Relays
10	Air temperature sensor	21	Starter
11	Butterfly valve box	22	Battery

General arrangement of the Fenix multipoint injection system for the R-25 GTX.

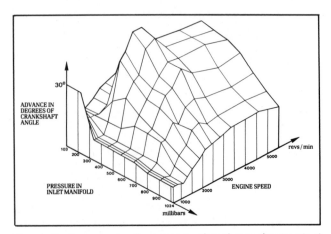

Spark-timing map for the same engine shows the greatest spark advance in the maximum-torque range, under high-load conditions.

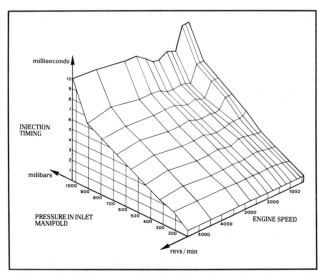

Fuel injection map for the Fenix-equipped 2.2 liter Renault engine shows a basically linear rise in fuel quantity when manifold pressure is increased.

injection without losing its benefits. Encouraged by a commitment from Renault, Bendix then began studies for a fuel injection system applicable to a compact base-level automobile.

That led to the adoption of the TBI (throttle body injection) concept, with a single injector, which could best be made to meet the stringent cost targets. Performance requirements were met with less difficulty than expected. Installed on the 1983 model Renault Alliance, the TBI gave excellent fuel economy and adequate performance, and it met the EPA emission control standards with a relatively small catalytic converter.

This system had digital control. The injector nozzle was mounted above the throttle plate, and metering was ensured by a microprocessor programmed

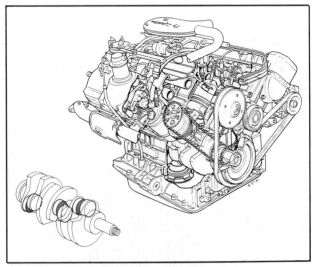

Fenix engine management system installed on the even-firing 90-degree V-6 of the R-25. It puts out 160 hp at 5400 rpm from 2849 cc on a 9.5:1 compression ratio (with knock detector).

with a closed-loop speed-density control strategy. The gasoline flow through the injector was proportional to the pulse width of the signal sent to the injector valve.

Apart from the single-point injector valve, the throttle body assembly included a fuel pressure regulator and an idle speed actuator. A low-pressure fuel pump was immersed in the tank. It was an electric motor-driven roller-vane pump, controlled by the microprocessor by means of a relay. Sensors reported on manifold pressure, ambient air temperature and coolant temperature. There was provision for adaptive controls such as idle speed stabilization and onboard diagnostics.

The Bendix TBI system proved extremely versatile in terms of programmability, which ensured easy adaptation to a variety of engine and vehicle sizes. It could also be easily modified to meet changed requirements for an existing installation.

The next step was what Bendix called IEC (Integrated Engine Control). Based on the TBI of the 1983 Alliance, IEC broke new ground in consolidating the sensors, the electronics and the fuel injector into the throttle body. Prototype systems were tested in the winter of 1983-84, with promising results.

The key to successful application, Bendix warned its potential clients, rested on the use of thick-film hybrid temperature sensor modules, which can be economically produced in high volume. Bendix had established that compared with other sensors, thick-film nickel elements were able to sense temperature with greater accuracy, faster response, longer life, and improved resistance to mechanical shock and chemical effects.

Thick-film applications had been under development since 1973, mainly for passive components such as resistors. In the IEC, Bendix used thick-film hybrid technology with unpackaged semiconductor chips. These chips were mounted on a ceramic substrate previously printed with a wiring pattern and were then wire-bonded to the circuit track. The modules were encased in a package that resisted well, providing rapid heat transfer and high vibration tolerance.

In the TBI with IEC, state-of-the-art film technology was used not only for new sensor designs, but also as part of the integration project that combined all electronics and sensors into the throttle body assembly. The TBI-IEC was produced by the Bendix Electronics Control Division, and it was adopted for the 1986 model Renault Alliance (R-9) and Encore (R-11) built by American Motors Corporation.

For European-market applications, the TBI-IEC was overtaken by the more advanced and highly versatile system known as Fenix. This system was developed in close collaboration between Renault and American Motors engine specialists and the electronics and software engineers of Renix. Fenix was revealed in May 1984 as a fully digital electronic engine control system. It was immediately adopted for

the Renault 25 GTX, equipped with the four-cylinder 2.2 liter engine, giving intermittent multipoint injection.

As in the original D-Jetronic and the Weber sequential injection system, fuel-metering calculations in the Renix system are made on the basis of continuous measurement and analysis of the relationship between intake manifold pressure and engine speed. Spark timing is also coordinated in accordance with speed-density and engine-knock input signals.

Renix chose a speed-density ratio as the basis for fuel-metering calculations, in preference to mass airflow measurement. Why? Mainly because of lower cost. Bendix has determined that the speed-density ratio method will give excellent transient response as long as the air temperature sensor is mounted in a good location. Its EGR (exhaust gas recirculation) strategy must also be carefully worked out for engines with open-loop systems.

The Motorola control unit calculates the amount of fuel mainly on the basis of engine speed, manifold absolute pressure and air temperature, with correc-

Renault 25 V-6 injection. The engine is mounted longitudinally ahead of the front-wheel axis, driving the front wheels.

tive factors such as coolant temperature, battery voltage, atmospheric pressure and crank angle coming into play in a secondary role. Switches to warn of wide-open throttle and fully closed throttle also send signals to the electronic control unit. Its output signals set the width of the injector drive pulse. All four injectors deliver at the same time, once during each crankshaft revolution.

The same control unit orders the spark timing and sends orders to the coil. The spark-advance curve is patterned on the injection map, with corrections for coolant temperature, ambient temperature, and sharp rise and drop rates in rpm.

A knock detector, mounted on the cylinder head, provides signals for each individual cylinder. The relative severity of these signals is evaluated before the spark is actually retarded. The evaluation takes only milliseconds, and the idea is to avoid needless spark retard, since the engine runs most efficiently right on the verge of knocking.

A single-point version of Fenix was developed, but has not yet come into application. Bendix (sole owner of Renix since Renault sold its share in a 1986 economy drive) also developed a sequential injection version of the multipoint Fenix, which was adopted for the 1987 Jeep Cherokee powered by the six-cylinder four-liter AMC engine. Finally, in September 1987, Renault switched from Bosch KE-Jetronic on the R-25 V-6 Injection to the Fenix intermittent multipoint system.

Brief adventure in Formula One

For the 1985 season, Renault switched from Kugelfischer to Renix electronic fuel injection for its 1500 cc V-6 turbo-intercooled Grand Prix racing car engines. This system was developed on the basis of Renix components by Renault's electronics research department.

Renault's Formula One engine was a twenty-four-valve V-6 with a speed range extending to 12,000 rpm. Thus, the engineers wanted twelve electromagnetic fuel injector nozzles (one per intake port) to satisfy the engine's need for fresh fuel in the minimum of time. The mechanical makeup of this engine

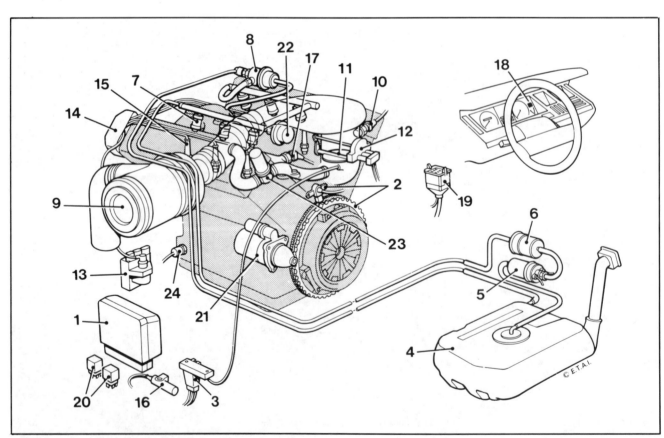

Schematic for the Fenix installation on the Renault V-6 engine. 1-Electronic control unit, 2-Crankshaft angle and rpm sensor, 3-Manifold pressure sensor, 4-Fuel tank, 5-Electric fuel pump, 6-Fuel filter, 7-Injector, 8-Fuel pressure regulator, 9-Air cleaner, 10-Air temperature sensor, 11-Throttle plate, 12-Throttle plate angle sensor, 13-Ignition module with coil, 14-Ignition distributor, 15-Spark plug, 16-Idle enrichment potentiometer, 17-Coolant temperature sensor, 18-Electronic malfunction warning light, 19-Diagnostic plug, 20-Relays, 21-Starter motor, 22-Fuel line pulsation damper, 23-Idle regulation valve, 24-Knock sensor.

can be compared to two half-engines, each with three cylinders in line, a separate inlet manifold and its own turbocharger mounted on its exhaust manifold.

The electronic control unit received a continuous flow of input data concerning the throttle plate angle (load), the crankshaft rotational speed (rpm) and the exact position of the crankshaft. Specific sensors supplied the control unit with signals indicating pressure and temperature in the intake manifold, but the rest of the engine sensor circuit from the passenger car was eliminated.

The same electronic brain also controlled the spark timing. In addition, the driver had a special gauge that gave him or her a visual impression of the main engine functions. Finally, a wireless emitter reported data on the engine's behavior to the pit crew, which could then prepare for remedial action ahead of time, in case of a malfunction.

The Deka-Series low-cost injector

Bendix introduced the Deka-Series injector in May 1986. It supplied a single-point version to Chrysler and a multipoint version to American Motors for use on fuel-injected 1987 models.

The Deka-Series came out of a well-defined research program aimed at creating a high-performance injector with a significantly lower cost than for existing systems. The conditions for putting it into production were that it be suitable for multipoint and single-point installations alike, fully interchangeable with older types of injectors and compatible with automated assembly.

Bendix engineers reasoned that taking the cost out of the injector could be accomplished only by wholesale simplification. In Detroit parlance, which is steeped in mass-production economics, that would mean combining several functions into a single part. But Bendix had taken that route earlier; this time, the opposite course was indicated.

Deka-Series injector by Bendix. This is a low-cost assembly that can help bring fuel injection into lower-priced cars.

The engineers split up the functions among a number of separate, inexpensive parts, whose combined cost was far lower than that of the multifunction parts they replaced. This method opened the door for using less expensive materials in some parts and reducing high-precision machining operations to the parts that required it.

The Deka-Series multipoint injector is a conventional solenoid valve, but instead of using the usual pintle-nozzle design, it delivers a pencil-beam spray. In contrast, the single-point injector sprays into the throttle body in a hollow cone pattern above the throttle plate.

The single-point nozzle works under relatively low pressure (10 to 20 psi), while the fuel delivered to the multipoint nozzle is pressurized at 35 to 90 psi.

One thing common to both versions is the metering method. Primary metering is ensured by a thin disc of stainless steel, with a precision orifice at the center. This thin-disc orifice technology gives the Deka-Series injector a tenfold increase in resistance to clogging, compared with rival injector types.

16

Bendix digital injection and Cadillac

Hardly had the analog control system gone into production than Bendix and Cadillac engineers were urging each other onward—to a digital control system. But there were many hurdles to cross, and it took some time before Cadillac was sure the system was ready for sale to the consumer.

The first production model Cadillac cars to be equipped with a Bendix digital control fuel injection system using central throttle body injection were the 1980 Eldorado and Seville. For 1981, the option was extended to all Cadillac cars with spark-ignition engines.

Bendix research engineers had the idea of using digital circuits in the electronic control unit long before Cadillac (and other prospective clients) suggested it should be tried. The company had applied for patents on digital control systems in 1970, 1971 and 1973.

The programming of a digital microprocessor requires serial computation as opposed to the parallel computations that take place in analog computers. Bendix's main problem with this concept came in two parts. The first lay in how to develop sensors that could supply analog signals from all the operating

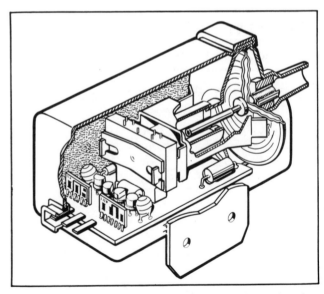

Manifold absolute pressure was chosen as the main criterion for fuel metering, because it reflects both speed and load variations. Signals from the vacuum sensor go through a linear variable differential transformer that feeds a flow of data to the control unit.

The Bendix digital fuel injection system adopted for the 1981 Cadillac V-8 included throttle body injectors and was combined with a Lambda-Sond exhaust feedback device.

parameters that have a bearing on air-fuel mixture needs. The second was how to find a way to handle the amount of computation needed for converting analog data to digital information in a minimum of time.

About 1972, Bendix began testing miniature semiconductor strain-gauge pressure sensors. It installed these sensors on cars having solid-state ignition systems, in which electronic signals from a breakerless distributor replaced the magnet-actuated reed switches.

By mid-1975, it was clear that Bendix would shortly be able to replace the hybrid analog system with a new design based on digital metal-oxide semiconductor technology.

This control unit concept would permit the elimination of the previous pulse-forming network and gating. It would substitute a standard printed circuit containing all circuit elements and networks, housed in a small metal case.

Such a microprocessor would operate from algorithms and the nonvolatile memories of metal-oxide semiconductors. Because of the high cost of such systems, important savings could be realized by combining other control functions, such as the ignition system, into a single integrated electronic system using the same control module.

This line of reasoning led Bendix to run two parallel research programs, one for developing the fuel injection system itself, and another for integrating spark timing, exhaust gas recirculation, supplementary air injection and other emission control systems with the fuel injection control module.

With fuel injection as the central pillar for a broad research program, the Bendix engineers discovered that the ideal injector valve operating frequency was once per crankshaft revolution. They also discovered that by varying the duration of the pulses applied to the injector valves *independent* of intake manifold pressure, significant improvements in efficiency could be made. Of course, that could not be accomplished without digital control.

Once a basic approach had been agreed on, Bendix modified the standard (hybrid analog) fuel injection system on a 1978 model 5.7 liter Cadillac V-8 to function with sequential multipoint injection instead of the two-group timed injection. The results were more than impressive: The exhaust analysis showed a forty-three percent improvement in carbon monoxide emissions, a twenty-eight percent improvement in unburned hydrocarbons and a seventeen percent improvement in oxides of nitrogen emissions.

Reaction from the Cadillac engineers was immediate and favorable. In their joint enthusiasm, Bendix and Cadillac worked hand in hand with the overoptimistic goal of getting a production model system ready for 1979 introduction.

Like any short-term project, this one had to be cost-accounted before any approval—and that almost gave the carburetor a new lease on life. The Cadillac engineers were told that the extra cost of sequentially timed injection would not be accepted. Grasping for the best lower-cost alternative, they removed the eight injector nozzles from the port areas and replaced them with just two injector valves enclosed in a cen-

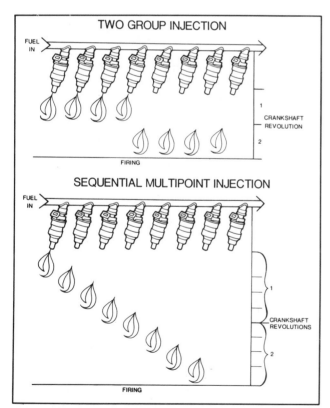

Bendix had been experimenting with sequential multipoint injection and had proved its advantages over the two-group injection method, but Cadillac chose further simplification and went to throttle body injectors.

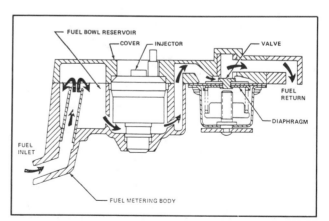

The fuel pressure regulator is an integral part of the throttle body. A relief valve operated by a diaphragm acts in response to atmospheric pressure to balance the fuel line pressure. A constant pressure drop is maintained across the injectors.

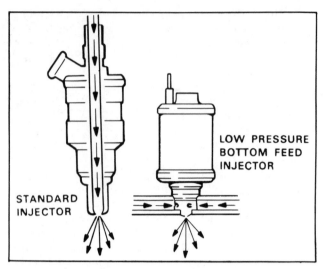

A new principle was adopted for the throttle body injectors: Fuel delivery was taken straight to the nozzle end, without going through the injector valve from the top.

tral throttle body that was mounted like a carburetor on top of the intake manifold.

This was an important deviation from earlier principles. It removed the mixture formation stage from the inlet ports to an area far upstream. From a purely technical point of view, this could be regarded as both good news and bad news. On one hand, the fuel was given much more time to atomize and mix with the air, which worked in favor of a more homogeneous mixture capable of lean-burn operation. But on the other hand, it introduced the risk of wetting

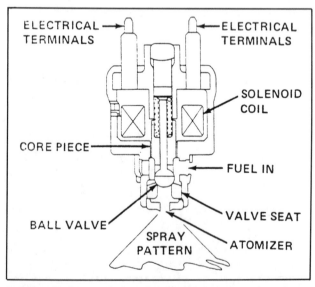

Schematic view of the Cadillac injector. The injector valves are electronically actuated. A solenoid valve opens the ball valve by raising the plunger, as in port-type injectors. The fuel is metered into the throttle body upstream of the throttle blades.

down the manifold walls with raw fuel during periods of mixture enrichment.

And so, as happens so often in Detroit, "good enough" replaced "the best."

In the system fitted on the 1980 Cadillac Eldorado and Seville, fuel was pumped from the tank and delivered to the throttle body after filtration. The pump was a twin-turbine type, electrically driven, built in a unit with the tank float system. Pump operation was controlled by the control module through a pump relay, going into action when the ignition was turned on.

The throttle body contained a pressure regulator, which maintained fuel-line pressure at a nominal 10.5 psi, and two fuel injector valves, which metered fuel into the airstream. The valves were positioned above the throttle plates, aiming the partly atomized fuel into the throats. The amount of fuel and the timing of the injection were dictated by the digital control module.

Why bother with timing for injectors mounted so far away from the cylinders? The answer is not self-evident, but becomes clear when you look at the airflow system.

The incoming air passed through a filter and then entered the throttle body, where fuel was injected. A special distribution skirt was part of the lower throttle body, directly below each injector valve, to ensure uniform distribution to all parts of the intake manifold. The timing was, in effect, unrelated to any single cylinder, but calculated to correspond exactly to the average mixture requirement of all the cylinders.

Throttle valves controlled the airflow rate in response to accelerator pedal movement. Idle speed was determined by the position of the throttle valves in combination with the idle speed control.

The throttle body was a cast-aluminum housing with two throats and two throttle blades mounted on a common shaft. One end of the shaft was connected to the accelerator linkage, and the other end to the throttle sensor.

The manifold absolute pressure sensor, the manifold air temperature sensor and the barometric pressure sensor gave the principal inputs serving to measure the amount of air entering the engine. An rpm signal was provided by the Delco High-Energy Ignition system.

On the basis of these inputs, the electronic control module made high-speed digital computations to determine the amount of fuel to be injected, on a continuous basis, with response in milliseconds. The pressure regulator was an integral part of the throttle body. It regulated the pressure by means of a diaphragm-operated relief valve which balanced fuel pressure to maintain a constant pressure drop across the injectors. The nominal fuel line pressure was maintained by the preload of a metal spring.

Any amount of fuel not needed to maintain the constant pressure drop was bled off into the return

Although it handles a host of extraneous functions, the control unit for the digital injection system is by far the smallest of the three shown here. At left, the analog control unit developed by Bendix in 1972-73. At right, an interim digital control design from 1976.

line to the tank. The quantity of fuel injected depended solely upon the valve-opening duration.

Both injector valves were electronically actuated on command from the control module. The valve body contained a solenoid whose core piece was a plunger, which was pulled upward by the solenoid coil when the coil was energized. Raising the plunger enabled the spring to push the ball-check valve off its seat, letting fuel begin to flow through the valve.

During cranking, both injectors were actuated simultaneously to provide a rich mixture for starting. In normal operation, the injectors were actuated alternately in time with each reference pulse from the ignition distributor. The intake manifold was a specific design for use with digital fuel injection, and it had a built-in exhaust heat crossover passage for the mixture enrichment heat riser valve.

In keeping with Bendix practice, fuel-metering calculations had two main factors: throttle plate position (load) and engine speed (rpm). The throttle position sensor was a variable resistor wire mounted on the throttle body and connected to the throttle valve shaft. The sensor continuously measured the angle of the throttle plates and sent an electrical signal to the control module, which processed the data for ordering mixture enrichment during acceleration and operating the idle control system.

Almost equally vital, the manifold absolute pressure sensor was similar to the unit used with the analog control module. It varied the pulse width of its electrical signal in accordance with changes in pressure. Intake air temperature was measured in the manifold.

The manifold air temperature sensor was a thermistor whose resistance changed as a function of temperature. The resistance went up when the temperature was low, and it fell off when the air temperature went higher. The coolant temperature sensor was also a thermistor, inserted just below the thermostat.

The electronic control module was a black box whose insides will forever remain a mystery to most of us. What we should know about it can be summed up in a description of its main functions.

The data sensors supply analog signals to the control module. The control module contains a battery of input/output devices that convert the input data into digital signals for the central processing unit, which can handle only digital information.

The central processing unit performs the mathematical calculations and logic functions required to produce the correct air-fuel mixture at all times. It also calculates spark timing and idle speed, and it commands the operation of emission control systems, powertrain diagnostics and cruise control.

Along with exhaust gas recirculation and auxiliary air injection, the control module has a capability for handling closed-loop feedback so as to instigate mixture variations based on analysis of the oxygen concentrations in the exhaust gas. Bendix began working on Lambda-Sond research in 1970 and had secured full patent coverage by 1974. The first car maker to use it was not Cadillac but Volvo, with Bosch-manufactured parts, on its 1977 240 series 2.1 liter four-cylinder models equipped with Bosch D-Jetronic fuel injection for sale in California.

17

General Motors fuel injection applications

Cadillac installed a Bendix digital-control throttle-body fuel injection system on the six-cylinder V-8 engine of the 1980 model Seville and Eldorado. The following year, this Bendix system became optional on all models available with the six-liter V-8. Lucrative as the contract was, it did not give rise to euphoria at Bendix, whose leaders realized full well they were just getting the benefits of a temporary bonanza.

Market projections made it clear that if they were remotely right about the pace of fuel injection installation in place of carburetors as original equipment, the volume would soon be so huge that General Motors

was unlikely to leave much room for an outside supplier. With all its special components divisions, General Motors was well placed to become its own supplier.

As early as 1978 a group led by John Schweikert at the GM Engineering Staff was then well into design and development work of production-model fuel systems, using components made up ad-hoc as well as parts from outside sources.

The lowest-cost proposition that emerged from this program was a single-point throttle-body fuel injection system, created by a team working under the direction of Lauren L. Bowler. Cadillac's senior staff engineer for fuel systems and emission control,

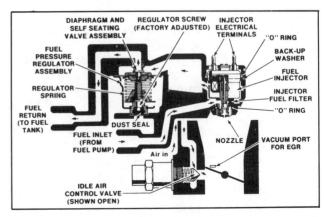

Throttle body with single-point injection was adopted for the Chevrolet Cavalier two-liter four-cylinder engine for 1983. This pushrod-ohv engine had a new Cyclonic swirl-port cylinder head and a 9.3:1 compression ratio.

Electronic single-point TBI for the 1982 Pontiac Iron Duke was responsible for four extra miles per gallon in highway driving, more reliable starting, unimpaired smoothness at all operating temperatures and higher power output.

Charles P. Bolles, made it the basis for a closed-loop digital fuel injection system with two electronically pulsed injectors in a single throttle body.

This system became standard equipment for the new 4.1 liter V-6 that replaced the six-liter V-8 in the 1981 Cadillac Seville and Eldorado. About 100,000 sets were made up before the end of the 1980 calendar year.

Rochester Products and Delco Electronics Division played central roles in the production of this system, which had a more advanced control module, capable of adjusting the air-fuel mixture eighty times per second. It controlled fuel metering, idle speed and spark timing.

This electronic control module received digital input from sensors monitoring the engine's operating variables such as manifold absolute pressure, engine coolant temperature, ambient barometric pressure, air-fuel mixture temperature, engine speed and crankshaft angle, throttle position, exhaust gas oxygen content and vehicle speed.

The microprocessor was programmed to use all the data as a basis for calculations enabling it to issue orders to the fuel injectors, idle air control motor, ignition distributor, torque-converter lockup clutch and other engine actuators.

Rpm and crankshaft angle signals were sent to the control module from the ignition distributor. An electronic switch, based on the Hall-effect phenomenon, was adopted to provide more accurate definition of the crank angle position.

The throttle body assembly contained the injector valves, a fuel pressure regulator, a throttle position sensor and an idle-air control motor. Fuel metering was based on speed and density measurements, using an AC-devised manifold absolute pressure sensor for control of the air/fuel ratio. Fuel entering the throttle body was channeled directly to the pressure regulator, which consisted of a diaphragm-operated relief valve to maintain a constant pressure drop across the injectors. Excess fuel went into a bypass and returned to the fuel tank.

The fuel injectors opened and closed as directed by signals from the electronic control module, and the duration of the opening time determined the amount of fuel delivered. Pulse duration varied between one and six milliseconds, depending on the engine's operating needs.

As a separate bonus, the system was capable of performing certain diagnostic functions. Because the electronic control module monitored the engine operating variables, it was able to identify and memorize malfunctions, alerting the driver by means of a "check engine" warning light on the instrument panel.

The 1983 model Cadillac 4.1 liter V-6 had its output raised from 127 hp at 4200 rpm to 135 hp at 4400, with a parallel rise in torque from 258 Nm at 2000 to 272 Nm at 2200 rpm.

TBI on Pontiac's Iron Duke

The engine known as the Iron Duke was a Pontiac project for a low-cost iron-block pushrod-ohv power unit to replace the ill-fated aluminum-block overhead-camshaft Chevrolet Vega engine. The other General Motors car divisions adopted it for their light cars, and for 1982, the Iron Duke was fitted with throttle-body injections.

The system included a single-barrel throttle-body assembly, mounted on the intake manifold in place of a carburetor, an electronic control module, a number of sensors, a tank-mounted electric fuel pump, an 89 mm ignition distributor with electronic timing control and a three-way single-bed beaded catalytic converter.

Pontiac's experience with fuel injection dates back to its use of the Rochester system in 1957. But the division's engineering staff also possessed more recent knowledge.

In 1969-70, when Hulki Aldekacti was head of advanced power systems, Pontiac was testing an overhead-camshaft hemi-head seven-liter V-8 with port-mounted fuel injectors and eight individual throttle valves on the intake runners. The fuel pump was driven by the ignition distributor, which was mounted horizontally on the front cover, and the fuel metering was combined with the spark timing. This engine was capable of 8000 rpm and more. On a 12.0:1 compression ratio, it was estimated to put out 650 hp

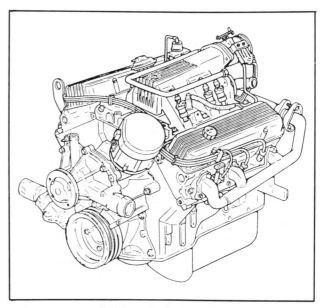

Multipoint fuel injection with Bosch fuel rails and injectors were adopted for the 1984 model Buick Century and Oldsmobile Cutlass Ciera with the 3.8 liter 90 degree V-6 engine. All six injectors sprayed once per crankshaft revolution, which meant that two successive injections were mixed with one dose of incoming air for each combustion cycle.

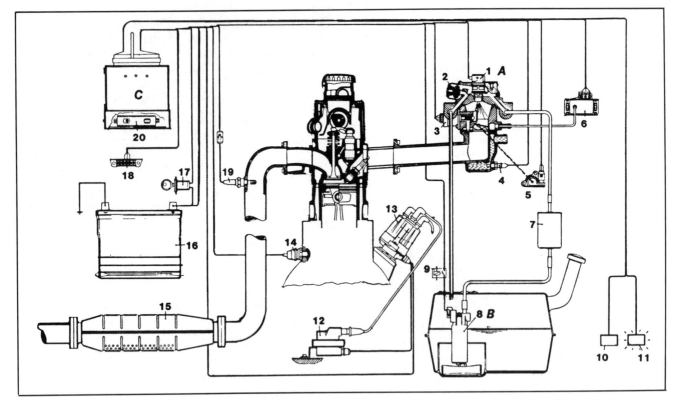

A-Throttle body, B-Fuel tank, C-Electronic control unit, 1-Injector valve, 2-Fuel pressure regulator, 3-Idling stepper motor, 4-Coolant temperature sensor, 5-Throttle plate potentiometer, 6-Manifold pressure sensor, 7-Fuel filter, 8-Fuel pump, 9-Fuel pump relay, 10-Vehicle speed sensor, 11-Engine warning light, 12-Ignition coil, 13-Ignition distributor, 14-Oil pressure switch, 15-Catalytic converter, 16-Battery, 17-Ignition key, 18-Diagnostic functions, 19-Oxygen sensor, 20-Program selector.

at 7500 rpm. But the project was abandoned before it ever left the engine lab.

The decision to put TBI on the 1982 Iron Duke owed nothing to this experiment, but was based on still newer technology. Anyway, it was a corporate decision, prompted by Chevrolet, and supported by Cadillac, Buick and Oldsmobile.

The 2.5 liter four-cylinder Pontiac engine was the base unit for the A-body cars (Oldsmobile Cutlass family), the X-body cars (Chevrolet Citation family) and the F-body cars (Camaro and Firebird).

The Rochester TBI was based on the same GM Engineering Staff studies that Cadillac had adopted. In contrast with the 4.1 liter Cadillac V-8, the Iron Duke was a mass-produced engine, and the closed-loop system was simplified as much as practicality, drivability, servicing and durability demands would allow.

The throttle body held one single injector valve, and the injector assembly received fuel under medium (32 psi) pressure, which was reduced to a stable 10 psi by the pressure regulator. An electric-impulse plunger on the injector opened for a conical fuel spray downward into the intake manifold above the throttle plate, according to pulses determined by the electronic con-

trol module. The opening times dictated the amount of fuel.

The electronic control module was mounted inside the passenger compartment to avoid the underhood environment with its high temperatures. The various sensors supplied updated general information ten times every second, and for critical data on emissions and drivability conditions, every 12.5 milliseconds.

Idle speed was controlled by supplementary air, bled into the throttle body below the throttle plate. An idle-air control motor functioned by moving a tapered pintle in an orifice to vary the amount of air that by-passed the throttle plate. A throttle-position potentiometer attached to the throttle plate shaft measured the plate's degree of rotation and provided a voltage that varied proportionally with the throttle plate angle.

Compared with the Cadillac V-6, there were several minor innovations. For instance, the oxygen sensor (Lambda-Sond) was contained in a cone-shaped zirconia ceramic body, mounted in the exhaust manifold upstream of the catalytic converter. Another point of difference was the adoption of a new electronically controlled ignition distributor which contained no conventional flyweights nor vacuum advance mechanism.

On an 8.2:1 compression ratio, the 2474 cc Iron Duke with TBI put out 90 hp at 4000 rpm, or an increase of 5 hp without running the engine faster. Peak torque was increased from 170 to 182 Nm at the unchanged crankshaft speed of 2400 rpm. This was really the start of things for GM. Rochester made 600,000 TBI sets in 1981 and 800,000 in 1982.

At the same time, GM began fitting the Rochester TBI on the 1.8 liter overhead-camshaft four-cylinder engine built in Brazil when installed in US-assembled J-body cars (Chevrolet Cavalier family).

With this engine, the 1983-model Pontiac 2000 became the first GM car to break the 50 mpg barrier in EPA ratings (highway mode). Chevrolet had also developed a Twin-TBI setup for the five-liter V-8 fitted in the Camaro and the 5.7 liter Corvette V-8. Pontiac adopted the five-liter V-8 for some Firebird models, starting in 1982.

The V-8 installation became known as "crossfire" injection because the throttle bodies were mounted above the banks opposite the ones they served, with the ducts crossing each other to provide adequate length. With a 9.5:1 compression ratio (and knock detector) the five-liter V-8 put out 165 hp at 4200 rpm with a peak torque of 326 Nm at 2400 rpm. That corresponds to a 20 hp gain over the carburetor version of the same engine (at 4200 rpm).

For 1983, General Motors began using TBI on the Chevrolet four-cylinder two-liter engine for the J-body cars. Since 1981, all Chevrolet gasoline engines came equipped with the CCC (Computer Command Control) system on the carburetor: It controlled spark timing as well as fuel metering, and in some automatic transmission cars managed the action of the torque-converter lockup clutch. The step to TBI was therefore not so radical as it might appear.

Multipoint injection on GM engines

Buick had perhaps neglected to pay attention to fuel injection in favor of concentrating on turbocharger applications. The first fuel-injected Buick was the 1982 Skylark, with the TBI-equipped Iron Duke supplied ready-to-go by Pontiac.

Buick's pioneer work on turbocharging brought the Flint-based division a lot of experience with electronic engine controls, however. In 1978, there were only three production cars in the world available with turbocharged engines: Saab, Porsche and Buick.

The 1981 model Buick Regal with the turbocharged 3.8 liter V-6 engine introduced a sort of electronic spark timing; the previous one acted on the carburetor.

Sensors for coolant temperature, manifold pressure and barometric pressure, detonation with corrective data from an oxygen sensor (Lambda-Sond) on the exhaust manifold, provided inputs enabling the electronic control modules to set the spark timing, adjust the air/fuel ratio and regulate the EGR on a continuous basis.

When Buick turbocharged the 4.1 liter V-6 on an experimental basis in 1983, Delco Electronics Division supplied a new on-board computer that handled all engine functions, including the turbocharger bypass valve, sequential port-type fuel injection, and water and methanol injection.

The fuel injection system was a mixture of Bosch and GM technology, using Bosch port-mounted injectors, GM electronics and a general operational scheme derived from the L-Jetronic. General Motors paid Bosch for the rights to use its patents, and purchased various Bosch components.

A very similar system became available as an option for the 1983 Oldsmobile Calais with the three-liter V-6 engine.

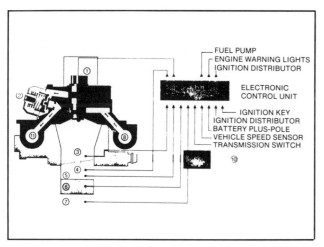

1-Injector valve, 2-Fuel pressure regulator, 3-Throttle plate potentiometer, 4-Idle air valve, 5-Manifold pressure sensor, 6-Coolant temperature sensor, 7-Oxygen sensor, 8-Fuel delivery line, 9-Fuel vapor canister, 10-Fuel tank vent, 11-Fuel return line.

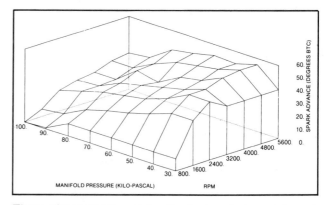

Three-dimensional map for the fuel metering and spark timing on the Multec-injected Opel engine. The system was based on a speed and density equation and required no mass airflow meter.

About that time, AC spark plug division went into production with a new mass airflow meter using twin sensors and averaging the readings. Buick used it on the 4.1 turbo V-6 and Chevrolet was planning to use it on its new 2.8 liter V-6 engine.

For 1984, Buick adapted its sequential port-type fuel injection to the turbocharged 3.8 liter V-6 engine of the Regal T as standard equipment. The same engine was offered as an option for the Riviera, later also for the Century Custom and Limited.

Cadillac adopted a closely similar fuel injection system for the new 4.1 liter aluminum-block V-8 that replaced the V-6 of the same displacement in the 1984 Seville and Eldorado. It put out 137 hp at 4400 rpm, reaching maximum torque (271 Nm) at half that speed.

What about the four-cylinder engines? Starting with the 1984 model 2000 SE, Pontiac was offering a turbocharged 1836 cc overhead-camshaft four-cylinder engine using a fairly straight Bosch L-Jetronic with GM electronics. In addition to the fuel metering, the electronic module also controlled spark timing and turbo-boost (bypass valve). This engine put out 150 hp at 5600 rpm and the torque curve was nearly flat—at 204 Nm—from 2800 to over 5000 rpm.

Rochester Products Division was ready with a multipoint simultaneous-injection system when 1985-model production began in the GM plants. Some applications were dependent on Bosch fuel rails and injectors, and all relied on AC spark plug and Delco electronics for essential components.

Cadillac replaced the two-liter TBI-equipped Chevrolet with a four-cylinder engine in the Cimarron with the new 2.8 liter multipoint port-injection V-6.

This was a Chevrolet engine, produced in three different packages: one for the Celebrity (A body), the second for the Citation (X body) and the third for the Cavalier (J body). With an 8.4:1 compression ratio, it put out 129 hp at 4800 rpm with a maximum torque of 216 Nm at 3600 rpm and found extensive use also in the A-, X- and J-body cars from Pontiac, Oldsmobile and Buick.

For the 1985 model year, Buick began manufacturing a new three-liter V-6 engine with Bosch rail-type sequential fuel injection and Bosch injectors, Delco electronics and the AC mass airflow sensor, Delco electronic vacuum regulator valve for the EGR and an AC high-pressure in-tank fuel pump. It also had computer-controlled coil ignition which elimi-

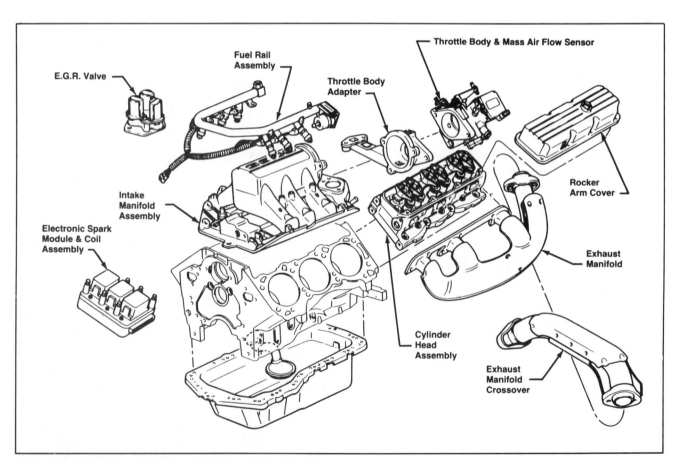

Exploded view of the 1988 model 3.8 liter 90 degree V-6 engine with sequential multipoint fuel injection.

nated the distributor, using a Magnavox system first introduced on the sequentially injected 3.8 liter turbo V-6 during the 1984 model year.

This engine had originally been planned for the 1986-model H-body cars (Le Sabre class), but instead was rushed through for the 1985 model N-body cars (Pontiac Grand Am; Oldsmobile Calais; Buick Somerset Regal). Its maximum output was 120 hp at 4400 rpm, with a peak torque of 204 Nm at 2000 rpm.

In a corollary move, Chevrolet replaced the "crossfire" twin-TBI injection setup on the 1985 Corvette V-8 with a new TPI (tuned port injection). It was basically a Bosch L-Jetronic with a hot wire airflow sensor and some variations provided courtesy of General Motors.

The incoming air collected in a plenum chamber mounted on top of the engine, with tuned runners curving out and under, to the ports. An eleven-percent average gain in fuel economy was obtained by putting the injectors in the intake ports. Maximum output was raised from 205 hp at 4000 to 230 hp at 4300 rpm, while the torque curve was shifted from a peak of 394 Nm at 2800 to 449 Nm at 3200 rpm.

Toward the end of 1985, Buick unveiled its Wildcat, an experimental vehicle powered by a four-camshaft twenty-four-valve 3.8 liter V-6 engine with sequential fuel injection, tuned to put out 230 hp at 6000 rpm, with a peak torque of 333 Nm at 4000 rpm. By that time, Buick was already in production with its 1986 model turbo-intercooled 3.8 liter V-6, using Bosch fuel rails and injectors. It also had a new high-pressure positive-displacement roller-vane fuel pump located inside the tank, providing a fuel-pressure range from 28 psi at idle to 51 psi at full boost. Intercooling boosted power output from 200 to 235 hp at 4400 rpm and maximum torque from 408 to 449 Nm at 2800 rpm.

In January 1986, Chevrolet announced an experimental high-performance prototype, the Corvette Indy, equipped with a fuel-injected thirty-two-valve twin-turbo intercooled four-camshaft 2.65 liter V-8 of about 600 hp. It had an induction system with sixteen runners, one for each intake valve, and sixteen Rochester Multec fuel injectors positioned behind the ports.

This engine design was directly derived from a CART-series racing engine by Ilmor Engineering of Brixworth in England for the Penske-March racing team.

Within months, however, Chevrolet showed a second version of the Corvette Indy, powered by a naturally aspirated V-8 of considerably bigger dimensions. This 350/32 engine was a 5.7 liter V-8 with four valves per cylinder and twin-cam heads on each bank.

It had narrow-angle pentroof combustion chambers, self-adjusting hydraulic valve lifters, chain drive to all camshafts and a hydraulic chain tensioner. With a fuel injection system closely based on the 2.65 liter

V-8 setup, it put out a less-than-expected 380 hp at 6000 rpm, with a no more jubilant peak torque of 503 Nm at 3800 rpm.

Yes, the engine was much closer to production specification than the pure-racing job, but Chevrolet had been getting the same specific power from pushrod-ohv sixteen-valve V-8s with carburetors twenty years earlier! Which meant there was lots of room for performance-oriented development in the power unit of the Corvette Indy.

The Multec injector was a new Rochester development with lots of promise. Mainly intended for port injection, it is also suitable for throttle-body (single-point) injection. It went into production in 1987 in the GM-Europe engine plant at Aspern, next to Vienna's airport.

The makers claim it provides faster response, improved fuel atomization and closer control over the spray pattern. It runs with low operating voltage and ensures clean plugs with any gasoline blend. What are its secrets?

Its construction is based on a new type of ball valve with an internal director plate which provides more sensitive operation. The ball valve opening interacts with a series of holes in the plate to ensure a fuel flow of 1.95 grams per second.

The metered fuel is forced through six holes in the director plate. These holes are not parallel, but drilled at a ten-degree angle from vertical, toward the center. When the injector is energized, fuel flows through the holes, merging in the center, from where it is deflected into a fifteen degree conical spray. Precision machining of the plate and the holes is required for the accurate spray pattern that is essential for the operational advantages claimed, and to make sure that

Olds Quad-4 Turbo prepared by Batten Engineering, with electronic fuel injection.

the director plate is insensitive to fuel properties and varnish build-up. Correctly machined, it eliminated varnish build-up by having a spray tip that shields the plate from fuel particles entering from the manifold.

Opel began using Multec injectors in four-cylinder engines for the Corsa and Kadett in 1987, but no installations were made on American-made GM cars until the 1988 model year. Several significant product improvements were, however, achieved in the meantime.

Rochester's carburetor production dropped from 4.4 million in 1980 to 1.8 million in 1986. For 1986, the output of the 3.8 liter V-6 with sequential multipoint injection for the Riviera (optional in the Electra and Park Avenue) was raised to 142 hp. Buick claimed that the change also brought improved response, smoother idling and lower fuel consumption. At the same time, Cadillac raised the output of the 4.1 liter V-8 from 127 to 132 hp at 4200 rpm, with a corresponding rise in torque from 258 to 272 at 2200 rpm.

On the 1986 Fiero GT, Pontiac went to Bosch-licensed GM multipoint injection with three-way catalytic converter oxygen sensor, EGR and exhaust-manifold air injection, for the 140 hp 2.8 liter V-6.

On the 1987 model Pontiac Iron Duke a dual-coil "direct fire" ignition system by Delco replaced the distributor type, and an improved TBI was fitted on a new intake manifold. Output increased to 97 hp (additional 6.6 percent).

In April 1987, a new Buick Regal with front-wheel drive appeared, as the first in the GM 10 program, with a 2.8 liter V-6 Chevrolet engine and Bosch-licensed multipoint fuel injection. Its maximum output was 127 hp at 4500 rpm with a peak torque of 217 Nm at 3500 rpm.

Production-model Oldsmobile Quad-4 engine, with multipoint injection and Multec port-mounted injectors.

Multec injectors and new engines for 1988

The single-point throttle-body injection system used on the four-cylinder two-liter and 2.5 liter engines was fitted with Multec injectors. Fuel was fed into the injector at 10.2 psi and injected into the manifold in a cone-shaped pattern at a five-degree angle. On a signal from the electronic control module an electromagnet opened the ball-and-seat valve at the base of the injector, thus releasing the fuel.

On the overhead-camshaft four-cylinder two-liter engine installed in the J-body cars (Cavalier family), the airflow measurement was replaced by a speed and density control system. The mass airflow sensor was eliminated and the manifold temperature sensor was moved to the air cleaner.

A manifold absolute pressure sensor determines the density of the incoming air and defines the loading condition; the manifold air temperature sensor, coolant temperature sensor and throttle plate angle potentiometer issue reference pulses that provide the corrective factors needed to compose the required fuel mixture for any engine speed. An exact rpm reading is taken from the ignition distributor.

Programmed into the electronic control module are volumetric efficiency tables with specific data for each vehicle induction and exhaust system combination at various EGR rates. These tables are compared with the continuous input data to compute the mass of air, which also gives a measure of the amount of oxygen entering the engine per crankshaft revolution.

The basic fuel metering calculation is further modified by feedback from the oxygen sensor (Lambda-Sond) with corrections to allow a margin for canister purge, to keep from exceeding stoichiometric mixture ratio during warmed-up light-load driving conditions.

Power enrichment is based on data from the throttle position sensor, and fuel supply is cut off on deceleration when a closed throttle position coincides with a certain drop in engine rpm (above a limit that is usually idle-speed plus thirty percent) for the prevailing ambient temperature. Cold-start enrichment is based on coolant temperature data and engine running time, while acceleration enrichment is based on the changes of throttle plate angle (and the rate of angular change).

Multec injectors were also used in the Bosch-rail sequential fuel injection system developed for the new ninety-degree 3.8 liter V-6 engine manufactured at Flint for all GM car divisions. The throttle body was built with an integral mass flow sensor, consisting of an air chamber that held a wire that was electrically heated to 75°C above ambient air temperature. It enabled the control module to calculate the total mass of intake air, since the amount of current required to maintain the 75°C temperature in the wire was proportional to the mass of air being ingested. Response to changes in airflow is immediate. And the sensor takes less than thirty milliseconds to compose and send a stabilized airflow reading to the electronic module.

Every 3.125 milliseconds the electronic control module carries out an updated airflow calculation and adjusts the fuel metering as required for a constant stoichiometric mixture. A 12 mm dual slope pintle nozzle was used for idle air control, with a pintle profile calculated to give the best idle characteristics at low ambient temperatures. EGR was ensured by a three-valve solenoid operated control valve. With a manifold vacuum of 355.6 mm of mercury and wide open throttle, it ensured a flow rate of 9.45 grams per second.

The port-mounted fuel injectors were repositioned for improved breathing and greater fuel efficiency, and the Delco coil-ignition system (without distributor) was adopted. Three electronically controlled coils fired every time a piston approached top dead center.

The digital electronics for the knock detector was a piezo-electric crystal connected to the control module like a sensor. At a predetermined knock intensity, the control module ordered the spark advance to be trimmed, immediately resuming a gradual advance until renewed knock was registered.

Compared with the earlier 3.8 liter V-6, the switch to sequential fuel injection and Multec injectors brought a ten-percent improvement in 0-60 mph acceleration; better than 1 mpg saved in average fuel consumption, lower exhaust emissions and smoother running throughout the speed range. Maximum output increased by ten percent to 165 hp at 4800 rpm, while the torque peak of 286 Nm at 2000 rpm represented a five-percent rise.

In adopting Multec injectors for the throttle-body injection system on the five-liter V-8, Chevrolet stuck closer to four-cylinder engine technology than to Buick's more advanced V-6. Among the innovations in the Chevrolet V-8 fuel system are new fuel-control algorithm to give better highway fuel economy and a new semi-closed loop system to enable the engine to run on leaner air/fuel ratios at highway speeds.

The control module was wired to monitor coolant temperature and spark timing, canister purge and constant vehicle speed (cruise control). Under certain conditions, the control module switches to an open-loop control mode, permitting operation on air/fuel ratios as lean as 16.5:1. Periodically, it switches back to closed-loop operation and runs the engine on a stoichiometric mixture for enough time to check all engine functions, and then returns to the open-loop operating mode.

Oldsmobile chose Multec injectors for its remarkable Quad-4 engine, which made its debut in the 1988 Calais GT. It's a sixteen-valve twin-cam four-cylinder

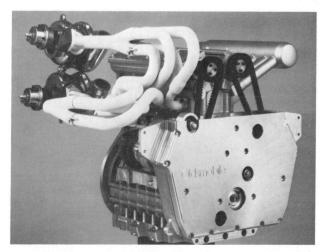

Olds Quad-4 twin-turbo prepared by Feuling Engineering, with all-mechanical fuel injection.

unit of 2.3 liters displacement, putting out 150 hp at 5200 rpm, with a peak torque of 218 Nm at 4000 rpm in standard form, on a 9.5:1 compression ratio. Each cylinder had four valves in a compact combustion chamber with sumped pistons. The centrally located extended-tip spark plugs of the Direct Fire ignition system were fired by dual coils housed in a die-cast cover which also acts as a heat sink. The coils fire each plug once every crankshaft revolution, without the aid of a distributor or spark plug wires.

Fuel metering relies on speed and density measurements, and port-type fuel injection is used, with a highly tuned aluminum intake manifold. A forward-mounted snorkel ducts the air to an inline cleaner element, through a zip tube and into the manifold's plenum, from where individual runners lead straight to the ports. Each Multec injector is located in an intake port at the best possible angle for optimum targeting of the fuel spray.

The production version of the Quad-4 is redlined at 7000 rpm, but Oldsmobile engaged two outside firms to explore its potential. One was developed by Batten Engineering of Romulus, Michigan, which reduced the displacement to two liters, turbo-intercooled it, and put in four injectors per cylinder: two for fuel, one for water, and one for other additives such as nitrous oxide. Rochester supplied its Multec electronic injectors, the fuel rail, tappets and throttle body.

The second engine went to Feuling Engineering of Ventura, California, where it was given a twin-turbo installation with intercooling and a purely mechanical fuel-injection system.

Both these special engines put out 1,000-plus hp at 9500 rpm. They were tested in the Aerotech experimental streamliner and clocked at speeds in excess of 400 km/h.

18

Chrysler's single-point injection

After their Electrojector experience, the engineers at Chrysler waited a long time before again tackling the problems of developing a simple, reliable and low-cost electronic injection system.

By virtue of its Aerospace Division in New Orleans and its Electronics Division in Huntsville, Alabama, Chrysler had been a leader in applying electronics to its own cars. Most notably, it had standardized electronic ignition on all car engines in 1972.

It was not until 1977, however, when Chrysler had arrived at the basic concept for a single-point continuous-flow injection system with electronic control, that a full-scale research program was started.

"We built twenty cars with electronic fuel injection in 1977, twenty more in 1978 and sixty-five in 1980," reported E. W. Meyer, Chrysler's chief engineer for motor/electrical. "We had to make sure we ironed out all the problems before we put the first system in a production car. We've tested the system under actual driving conditions at the Chelsea Proving Grounds, in Michigan, for hot-weather conditions, and in northern Ontario for extreme cold. In all, we have put over a million miles on these cars, and the results have been excellent—in reduced emissions, good fuel economy and outstanding drivability."

The Chrysler system was made standard on the 5.2 liter V-8 of the 1981 Imperial as the first step toward across-the-board availability.

The system offers complete electronic engine control, with the same control module commanding the spark advance. It monitors the air-fuel ratio electronically, compares it with an ideal ratio and adjusts it automatically to changing environmental and engine conditions. Chrysler holds or has applied for twenty-four separate patents covering almost every part of the control system.

The most obvious innovation in the Chrysler system is the electronic metering of mass airflow as well as fuel quantity. In addition to giving higher precision, electronic measurement minimizes the effects of

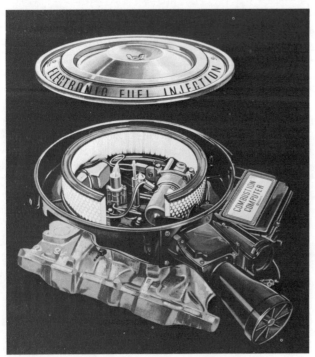

The throttle body assembly is centered in the air cleaner, which carries the electronic control unit.

manufacturing tolerances and wear on the system's mechanical components.

The Chrysler system maintains the quality of the air-fuel mixture by arranging the throttle blades and bore in geometrical relationships that make the inducted air shear, entrain and distribute the fuel evenly to each cylinder. In addition, the system electronically establishes the base air-fuel ratio for each individual car. There is no need for a manifold absolute pressure sensor.

Chrysler's system consists of three major assemblies—each one a functionally complete unit that can be tested separately. The fuel supply assembly is located inside the fuel tank. In addition to the conventional equipment that delivers fuel to the engine, this system also has an electric turbine pump and several check valves.

The second major assembly is the air cleaner, which also contains the airflow sensor, and the metering and ignition electronic module. Third comes the throttle body and mixture control assembly. This includes the control pump and its power electronics, the fuel-flow sensor, the pressure-regulating valves, the spray bars and the automatic idle speed motor.

The computer receives input data on three separate functions: the flow of air into the engine, the flow of fuel and the oxygen content in the exhaust gas. It compares these signals with an ideal calibration. When any of the signals is different from the calibration, the computer signals the control pump motor to deliver more or less fuel, depending on whether the mixture is too rich or too lean.

The pump in the tank feeds the fuel to the control pump. The control pump takes a small portion of the fuel and delivers it at a pressure of 21 psi at idle through the fuel-flow sensor and the low-pressure regulator valve, into the spray bar and then to the

nozzles. The control pump is a positive-displacement, slipper-type pump driven by a variable-speed direct-current motor.

Since the size of the openings for fuel is fixed, the pressure must be varied to deliver more fuel at higher speeds. At maximum speeds, the control pump delivers fuel at pressures up to 60 psi. The spray bar is designed to produce an even flow at low speeds. At higher speeds, a second spray bar automatically opens to deliver the full amount of fuel at the correct pressure. Fuel that is not needed to maintain a given speed is automatically returned to the tank through a low-pressure regulator valve and a return line.

The nozzles are connected to both the light-load and power spraybars. A number of airfoil-shaped nozzles are arranged in a circle around the injector body to refine the light-load fuel spray pattern. The light-load circuit supplies all the fuel the engine needs up to a metering pressure of 34 psi.

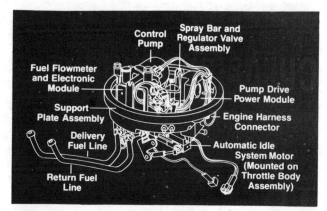

The fuel control subsystem rides on top of the throttle body and controls fuel metering on commands from the combustion computer.

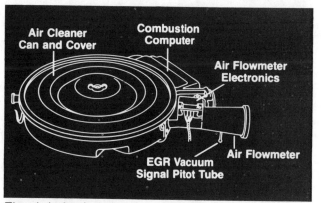

The air induction system receives air from two sources: unheated air enters from the fresh air inlet, and heated air comes from the heat stove on the exhaust manifold. The two are mixed in proportions controlled by a vacuum actuator, which moves a damper that admits a greater or smaller amount of heated air.

The fuel supply subsystem has an open circuit to the tank, returning all excess fuel to the tank from the control pump. The delivery pump is located inside the tank. (Note: On the diagram, In-take pump is misprint for In-tank pump.)

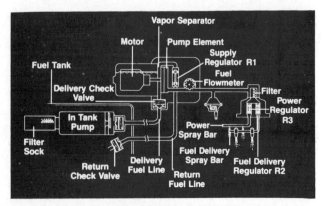

Fuel supply system schematic. The flow meter has a free-spinning wheel, whose rotational rate is proportional to the fuel flow. The turning vanes interrupt the light path between a light-emitting diode and a phototransistor, and signals on flow rate are processed by the flow meter module for passing on to the combustion computer.

At higher pressure, the power circuit goes into action, while the light-load circuit continues to spray at maximum. The power injector bar adds its own spray from a single larger orifice.

The airflow meter is a completely new type. It consists of a large-diameter venturi, mounted in close proximity to the air cleaner outlet, upstream from the throttle.

A ring of fixed, radially arranged vanes near the mouth of the venturi deflects the air into a clockwise swirl pattern. Owing to the centrifugal effects of this swirling air, the air pressure is lower at the center of the vortex than at the outside.

This vortex looks like a miniature tornado, with an eye that resembles a finely coiled thread, stretching throughout the length of the duct. Nearest the inlet end, the eye is centered in the duct. As it approaches the outlet end, the vortex expands to fill the widening

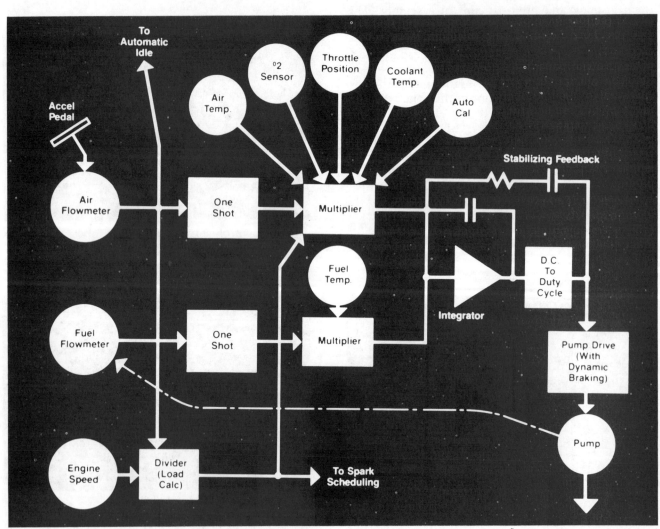

Inputs for fuel metering are shown in this schematic. The control pump is a positive-displacement, slipper-type pump driven by a variable-speed electric motor. The control unit controls the air-fuel ratio by ordering changes in the pump speed.

duct and becomes unstable, and the eye goes into an orbit around the center. What matters about this displacement of the eye is not the distance from the center or to the walls of the duct, but the frequency of its orbit.

This frequency is proportional to the volume of air passing through the duct per unit of time. Measuring the frequency of pressure variations along the orbit, therefore, provides an analog reading for mass airflow.

This measurement is taken by a row of U-shaped pressure-probe standpipes located near the outlet from the venturi. A silicon chip converts the low-pressure differential readings into a high-powered electrical signal. As the volume of air increases, the signal pulses faster. The pulse frequency tells the computer how much air goes into the engine at any given instant.

The fuel-flow meter consists of a cylindrical cavity containing a free-turning wheel. This wheel

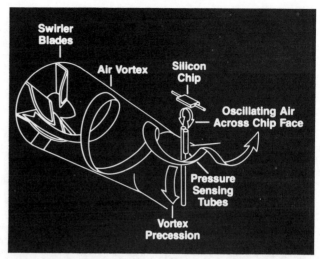

The airflow sensor works by having a U-shaped probe keep track of the movement of the low-pressure "eye" in the vortex.

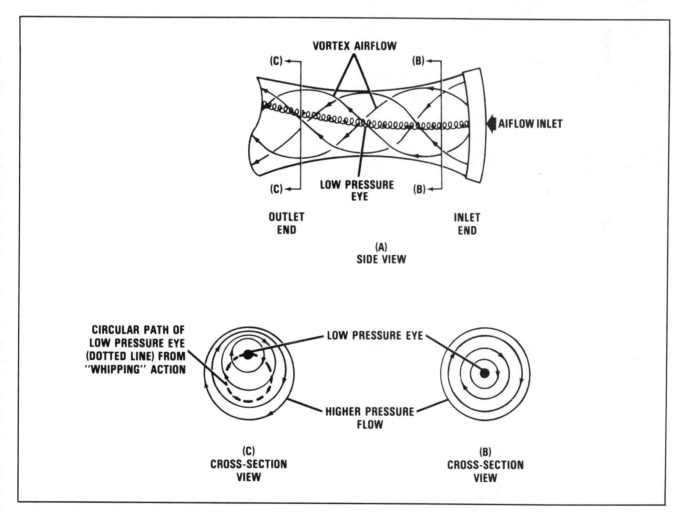

Vortex airflow patterns vary with airflow mass. The frequency of eye movement is proportional to the volume of air passing through the duct in a given time.

resembles a gear, but its teeth are actually small, curved vanes. It works like a paddle wheel. Fuel enters the cavity on a tangent and its pressure is brought to bear against the vanes. This leads to rotation in the wheel at a rate that is proportional to the fuel flow. The greater the flow, the faster the wheel spins.

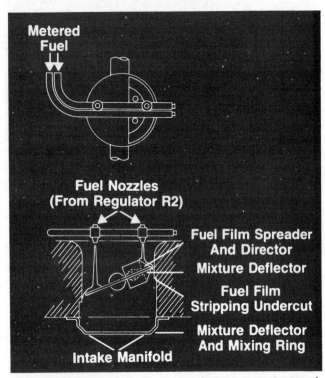

Fuel mixing takes place in an area that starts upstream of the throttle body, where the fuel bars emerge from the low-flow and high-flow control valves, and reaches through the manifold to the intake valves.

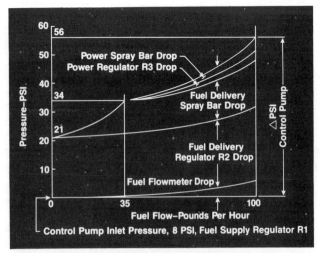

Fuel flow is related to control pump pressure in this chart, with curves showing the pressure drop occurring in the fuel flow meter, regulators and spray bars.

On one side of the paddle wheel is a light-emitting diode, and on the other side is a phototransistor. As the paddle wheel spins, it interrupts the flow of light between the two and sets up a signal that pulses faster as more fuel flows by. These pulses are relayed to the computer, and their frequency tells the computer how fast the fuel is flowing.

To the engine, the information that really matters is not the volume of air or fuel flow. The air-fuel ratio is what counts, not the volumetric ratio. Therefore, the system incorporates three additional sensors that translate the volume readings of air and fuel to mass readings.

Recall that the control module compares the air/fuel ratio with an *ideal* air/fuel ratio. This does not mean that Chrysler's fuel injection system is restricted to operation with a stoichiometric air-fuel ratio only.

E. W. Meyer explains: "We designed this system to operate on a very lean mixture of fuel and air—as little as one part fuel to twenty-one parts air. To operate the engine on a mixture this lean, you need three things: first, a uniform distribution of air and fuel to each cylinder; second, an even supply of air and fuel to a cylinder from cycle to cycle; and third, fuel preparation which gives the lowest specific fuel consumption and minimum specific oxides of nitrogen emissions, while at the same time controlling specific hydrocarbons output."

Basically, the signals from the two flow meters (air and fuel) trigger one-shot pulses of opposite polarities—positive and negative. These signals are fed into an integrator, which controls the pump motor.

An automatic calibration system individually establishes the air-fuel ratio on each car and compensates for changes in barometric pressure. There is a temperature sensor on the air cleaner housing so the combustion computer can adjust for changes in temperature that affect the density of the air. An oxygen sensor in the exhaust system tells the computer when the air-fuel mixture is too rich or too lean.

The fuel flow also makes contact with a silicon thermistor, which senses the temperature of the fuel. Its electrical conductivity varies in proportion with the fuel temperature, and it is thus able to send fuel temperature signals to the computer.

The mixture control device is a conventional two-part, single-throttle-shaft design like the air horn on a carburetor—with several important differences:

1. All the fuel is introduced above the throttle blade.

2. There are two fuel distribution bars over each port: a primary bar for low fuel flow and a power bar for staged high fuel flow.

3. The conventional butterfly throttle blade has a fuel-film spreader mounted above the blade and a mixture deflector mounted below.

The fuel streams impinge against the film spreader, which distributes the gasoline outward into thin films. The lower-velocity air at the larger blade openings draws the fuel film toward the sharp edges of the blade, resulting in a highly atomized mixture of fuel and air. The lower mixture deflector improves the cylinder-to-cylinder mixture uniformity.

A high degree of fuel atomization is vital for efficient and uniform combustion. A spherical droplet of fuel has a surface area proportional to the square of the diameter and a volume proportional to the cube of the diameter. Smaller and smaller atomized fuel droplets are vaporized in less time for an equal amount of heat.

Finely atomized fuel droplets of ten to fifty microns seem to be the right size. They mix more easily with the intake air and remain in the manifold longer. The action of the air shearing the fuel off sharp edges seems to be a very simple and effective device to achieve these droplet sizes.

A fuel pressure switch is included in the fuel control circuit between the flow meter and the injection assembly. Normally this switch is open, which indicates that the system has sufficient pressure for starting the engine. The valve is closed when the pressure is too low, and its closing completes a bypass circuit that drives the control pump at full speed until adequate pressure is restored.

When the engine is cranking, idling or running under a light load, only the primary fuel bar nozzles are delivering fuel. As speed increases, and the fuel flow reaches about thirty-five pounds per hour, the power regulator valve opens and fuel starts to flow from the power orifices.

The computer is a common center for four separate electronic circuits. The first is the fuel injection circuit, which gives speed orders to the control pump and thereby provides the basic factor for control of the air-fuel ratio. The second is the automatic calibration circuit, which monitors the oxygen feedback from the exhaust gas and serves to fine-tune the first circuit. The third is the electronic spark-advance circuit, which provides the initial ignition current and spark timing. The fourth is the automatic idle speed circuit, which goes into action whenever the throttle is closed and stabilizes the idle.

The initial production version of Chrysler's throttle body injection system proved so satisfactory that little or no modification was undertaken as the number of its applications spread throughout the corporation's engine range.

When the 1986 models were prepared, Chrysler developed a reduced-pressure version for the 2.2 and 2.5 liter four-cylinder engines available in the K-cars. The low-pressure (15 psi) fuel regulator was included in the throttle body, which also contained the throttle mechanism and throttle switch, single injection nozzle, and idle speed controller. This was a low-profile assembly for mounting directly on the intake manifold.

The injector was supplied by Bosch. It was a type that sprayed a hollow cone of atomized fuel particles into the central throat in a forty-five-degree pattern.

A new speed compensator was added to the idle control to ensure smoother and more consistent idling. It automatically adjusted the fuel metering in accordance with changes in power demand, such as the on/off cycling of the air-conditioning compressor.

By using lower pressure, Chrysler was able to fit a quieter in-tank fuel pump.

For 1988, Chrysler extended the use of its reduced-pressure injection system to the 3.9 liter V-6 and 5.2 liter V-8 engines installed in Dodge Ram pickup trucks and Ramcharger vehicles.

Compared with the 3.7 liter carbureted six-cylinder engine it replaced, the 3.9 liter V-6 offered a thirty percent rise in horsepower, a twenty-five percent gain in torque and a three percent fuel savings. Discarding the carburetor on the 5.2 liter V-8 in favor of fuel injection led to a twenty percent rise in horsepower.

The injection system used on these engines differs from that used on the four-cylinder passenger car engines in having a throttle body with dual throats, each with its own injector nozzle.

19

Ford style of fuel injection

Stuart Hilborn's fuel injection was fitted on the 1965 version of Ford's four-cam V-8 for the Indy 500, and some of these engines were later converted to Bendix injection. In 1968, Ford's Formula Two Cosworth racing engine was equipped with the Lucas Mark II fuel injection system. This was also adapted to the Cosworth V-8 Formula One engine a year later.

Aside from these racing experiments, Ford (both in Europe and in the United States) took a standoffish attitude toward fuel injection. Though a small number of Lucas-injected sixteen-valve four-cylinder Lotuses were fitted in Ford Escorts back in 1970, this did not lead to any supply contracts for Lucas.

When Ford of Europe embraced fuel injection in a serious manner, it was late: autumn 1982. That system was the Bosch K-Jetronic, adopted for the Escort XR3 i.

In 1988, Ford of Europe also had in production an Escort RS Turbo with KE-Jetronic, a Sierra 2.0 with L-Jetronic, a Sierra 2.6 with LE-Jetronic, a 204 hp Sierra RS Cosworth two-liter turbo with Marelli/Weber sequential fuel injection, and Scorpio models fitted with 2.4 and 2.9 liter versions of a new V-6 engine with Bosch LE-Jetronic.

Ford of Dearborn was in the field earlier, but was driven by different motives. In the United States, the tightening emission control standards focused attention on electronic engine management.

The 4.9 liter V-8 of the 1978 Lincoln Versailles was equipped with electronic controls including an exhaust gas oxygen sensor that could "feed back" signals to an electronically controlled carburetor. Known by the code name EEC-I, this system was the auto industry's first interactive electronic engine control system.

For some 1979 models, the two concepts were married into a second-generation system called the EEC-II. Those models were full-size Ford cars powered by the 5.8 liter V-8, which were for sale in California, and the full-size Mercury with the same engine, which was available in all fifty states.

A third generation of engine electronics, the EEC-III, was introduced for the 1980 Lincoln Continental and Continental Mark VI, powered by the five-

Ford's 1979 model electronic fuel injection system was a single-point type with two injectors mounted in the throttle body. A microprocessor controlled fuel metering, spark timing and various other engine functions.

liter V-8 engine in combination with Ford's automatic overdrive transmission.

With the arrival of the EEC-III, Ford jumped the barrier between the electronically operated carburetor and the electronically controlled single-point pulse-time-modulated injection system. After an essential field-testing period, the EEC-III was extended to the 1981 Ford LTD and Mercury Marquis.

EEC-I

The EEC-I system was built up around a completely solid-state module using a digital microprocessor and other custom-designed integrated circuits. Seven sensors were used to determine crankshaft position, throttle position, coolant temperature, inlet air temperature, manifold absolute pressure, barometric pressure and exhaust gas valve position. Using this information, the module calculated spark advance and the exhaust gas recirculation flow rate.

EEC-II

The EEC-II system was essentially a combination of the EEC-I with a three-way catalytic converter using oxygen feedback control. It was less complex, more reliable and lighter than the separate systems used earlier.

The first step in the development of the EEC-II's computer circuit was created by the engineers in Ford's EEC department to define the system requirements. The technical specifications were turned over to the department's semiconductor suppliers for integrated circuit design and fabrication. Through a complex process, the circuit was reduced to its final size: a chip about one-quarter-inch square. The EEC-II system had six chips, and each one contained 10,000 to 15,000 electronic devices.

Henry A. Nickol, then Ford's chief powertrain engineer, explained what was achieved in going from the first to the second generation of EEC systems: "We added capabilities to the system, but were able to reduce the package size by forty percent, the number of computer parts by thirty percent, and the weight by almost fifty-five percent. The system complexity was reduced and the costs were lowered dramatically. A third generation of the EEC system, to be introduced on 1980 cars, will be even less complex, but will provide more control functions."

In addition to controlling the major engine functions, EEC-II provided many other benefits. One benefit was that it controlled the purging of vapors in the storage canister used with the fuel-evaporative emission control system. Under certain conditions,

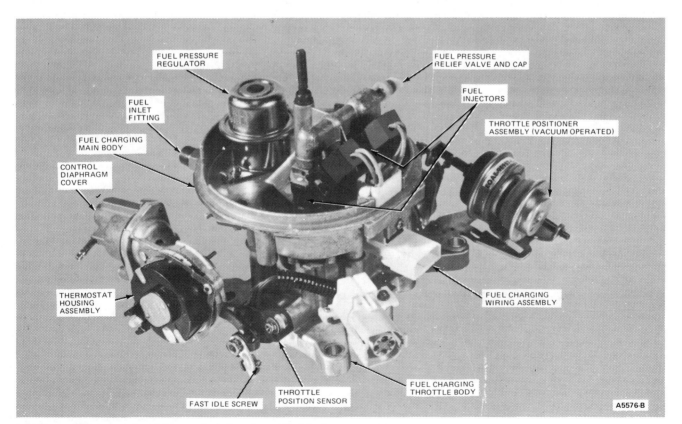

Injector valves are solenoid-operated, the amount of fuel injected varying only with valve-opening duration, as ordered by the microprocessor.

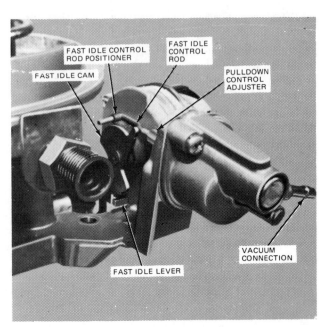

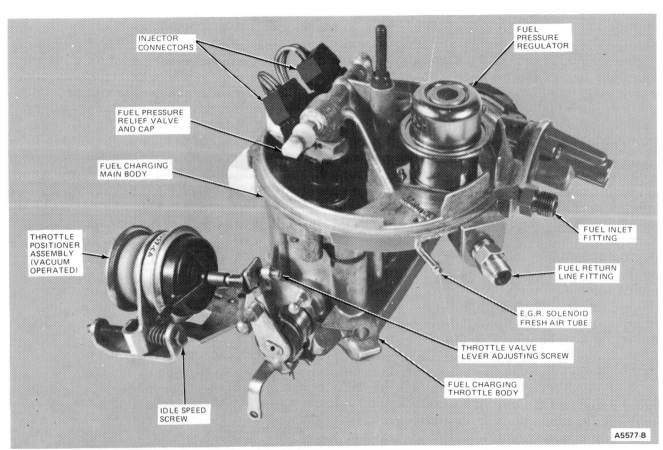

FAST IDLE CONTROL
ROD POSITIONER

FAST IDLE CAM

FAST IDLE
CONTROL
ROD

PULLDOWN
CONTROL
ADJUSTER

VACUUM
CONNECTION

FAST IDLE LEVER

Close-up view of fast-idle mechanism. Control diaphragm is located inside its cover (with vacuum connection at the end).

the purging caused the carburetor to run rich. When that happened, however, the EEC-II system simply sensed this change through the exhaust gas sensor and signaled the proper adjustment to the carburetor.

Another benefit was that the EEC-II minimized variations in idle speed through a throttle-idle positioner. For example, a sensor indicated whether the vehicle's air-conditioning system was operating. If it was, the control module signaled a throttle position solenoid to increase the throttle opening, compensating for the increased engine load while maintaining an acceptable idle speed. If the air conditioner was not being operated, the throttle-idle opening was reduced and thereby saved fuel. The throttle-idle positioner made similar corrections for cold-engine conditions and for high altitude.

The EEC-II system eliminated the need for special calibrations in high-altitude areas. In addition to the idle speed adjustment, the system sensed barometric pressure and made the necessary engine adjustments automatically.

The system also controlled the Thermactor (a pump that adds air to the exhaust system) by diverting

INJECTOR
CONNECTORS

FUEL
PRESSURE
REGULATOR

FUEL PRESSURE
RELIEF VALVE
AND CAP

FUEL CHARGING
MAIN BODY

THROTTLE
POSITIONER
ASSEMBLY
(VACUUM
OPERATED)

FUEL INLET
FITTING

FUEL RETURN
LINE FITTING

E.G.R. SOLENOID
FRESH AIR TUBE

THROTTLE VALVE
LEVER ADJUSTING SCREW

FUEL CHARGING
THROTTLE BODY

IDLE SPEED
SCREW

A5577-B

Mechanically, Ford's electronic fuel injection is far simpler than a carburetor, but it has numerous electrical connections that are not used on conventional carburetors.

Throttle position is set by vacuum—not by mechanical linkage.

this secondary air to either the catalyst or the exhaust manifold as required to maximize emission control.

EEC-II controlled idle mixture and initial spark automatically, thus eliminating the need to make these maintenance adjustments.

A Model 7200 "feedback" carburetor as used in the EEC-II system was the same variable-venturi carburetor introduced on the 5.0 liter V-8 engines in California in 1977, except for modifications that were necessary to include the feedback feature. One major modification was the addition of a stepper motor that adjusted the position of a rod in the carburetor to trim the airflow—rich or lean—required to maintain the desired ratio.

EEC-III

The essential difference between the EEC-III and the EEC-II lay in the use of single-point electronic fuel injection on the newer system. The control system, including its sensors and outputs, was basically the same.

Ford's electronic throttle body used with the EEC-III was produced at the Rawsonville, Michigan, plant of the Electrical and Electronics Division, where most of the company's conventional carburetors are also manufactured.

The microprocessors for the EEC-II system were supplied by Shibaura Electric Company (Toshiba) and the Electrical and Electronics Division of Ford Motor Company. Microprocessors for the EEC-III system were produced by Motorola.

National Semiconductor Company and Signetics supplied small integrated circuits for the EEC-II. Suppliers for the EEC-III included Intel (large-scale integrated circuits), National Semiconductor Company (small-scale integrated circuits and transistors) and Fairchild Semiconductor Company (small-scale integrated circuits).

At the heart of this electronic fuel injection system, Ford placed two electrically activated fuel-metering valves. When energized, these valves sprayed a precise quantity of fuel into the engine's intake airstream.

Both injector nozzles were mounted vertically above the throttle plates and connected in parallel with the fuel pressure regulator. The injector valve bodies consisted of a solenoid-actuated pintle and needle valve assembly. An electric control signal from the EEC-III electronic processor activated the solenoid. This caused the pintle to move inward off its seat, which allowed fuel to flow. The injector flow orifice was fixed and the fuel supply pressure was constant; therefore, fuel flow to the engine was controlled by how long the solenoid was energized.

The controlled high fuel pressure in the injectors, along with the precise change in fuel volume determined by the EEC-III, improved fuel distribution to

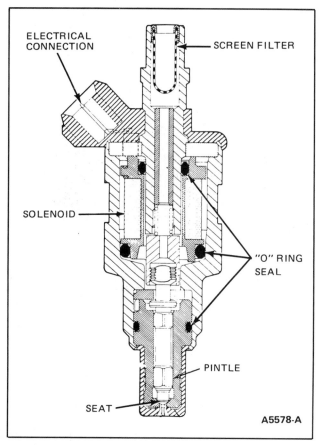

Ford's injector uses a pintle-type nozzle that delivers a fine fuel spray into the throttle passage. Fuel is fed in under constant pressure, and the injection is timed electrically. Quantity depends on opening duration.

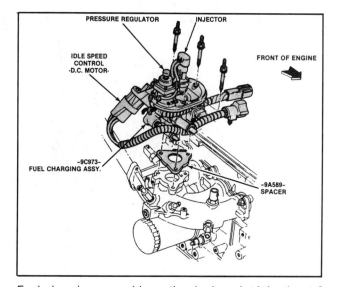

Fuel-charging assembly on the single-point injection 1.9 liter four-cylinder engine. Fuel is supplied under 30 psi pressure from an electric pump mounted inside the tank. The nozzle is mounted above the throttle plate, connected in series with the fuel pressure regulator.

each cylinder, compared with the carburetor it replaced. This in turn provided optimum drivability and economy while exhaust emissions were maintained at acceptable levels.

Fuel to the injectors was provided by a high-pressure electric pump mounted inside the fuel tank. A special primary fuel filter was located in the fuel line, under the passenger compartment. A smaller, secondary filter was installed in the engine compartment.

A fuel pressure regulator mounted on the throttle body, just ahead of the injectors, controlled fuel pressure to a precise and constant value of 39 psi. Excess fuel supplied from the pump but not needed by the engine was returned to the fuel tank by way of a fuel return line.

The pressure regulator was mounted on the fuel-charging main body near the rear of the air horn surface. The regulator was located so as to nullify the effects of the supply line pressure drops. It was designed so that it was not sensitive to back pressure in the return line to the tank.

A second function of the pressure regulator was to maintain fuel supply pressure following engine and

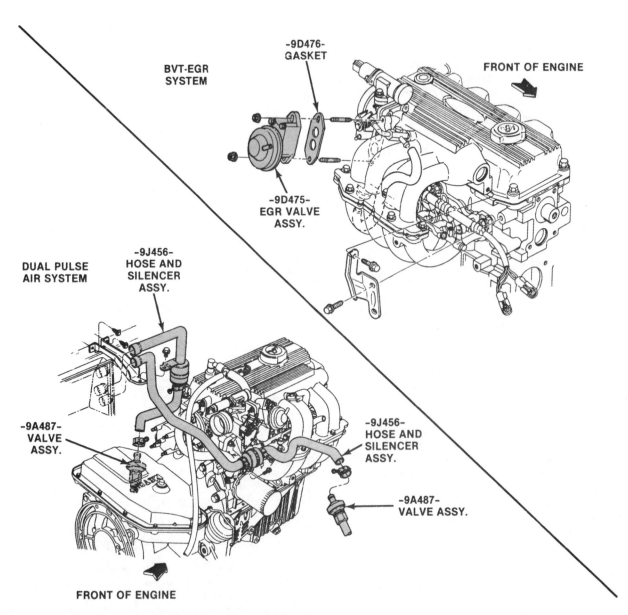

Multipoint version of the 1988 Ford 1.9 liter fuel injection engine, whose emission control systems include EGR by BVT (back pressure variable transducer) and a Thermactor dual-pulse supplementary air system.

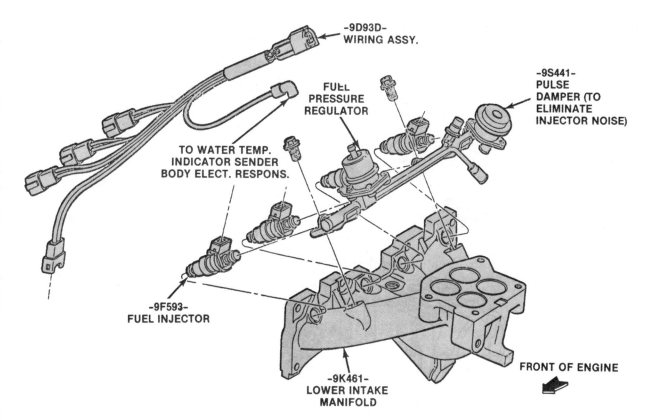

-9D93D-
WIRING ASSY.

FUEL
PRESSURE
REGULATOR

-9S441-
PULSE
DAMPER (TO
ELIMINATE
INJECTOR NOISE)

TO WATER TEMP.
INDICATOR SENDER
BODY ELECT. RESPONS.

-9F593-
FUEL INJECTOR

-9K461-
LOWER INTAKE
MANIFOLD

FRONT OF ENGINE

Multipoint electronic fuel injection on the 1988 model 2.3 liter four-cylinder Ford engine is controlled by the EEC-IV module. Fuel pressure for the turbocharged version is controlled by a vacuum-operated pressure regulator.

fuel pump shutdown. The regulator functioned as a downstream check valve and trapped the fuel between itself and the fuel pump. The maintenance of fuel pressure after engine shutdown prevented the formation of fuel line vapor and allowed for rapid restarts and stable idle operation.

Airflow to the engine was controlled by two butterfly valves mounted in a two-piece, diecast aluminum housing called the throttle body. The butterfly valves were identical in configuration to the throttle plates of a carburetor and were actuated by a similar linkage and pedal cable arrangement.

A throttle position (TP) sensor was mounted on the throttle shaft on the side of the fuel-charging assembly and was used to supply a voltage output proportional to the change in the throttle plate position. The TP sensor was used by the computer to determine the operation mode (closed throttle, part throttle and wide-open throttle) for selection of the proper fuel mixture, spark and EGR at selected driving modes.

For fast idle before the engine reached normal operating temperature, Ford used a throttle-stop cam positioner of the same type used on carburetors. The cam was positioned by a bimetal spring and an electric

positive-temperature heating element. The electrical source for the heating element was 7.3 volts from the alternator stator, which provided voltage only when the engine was running.

The heating element was designed to provide the necessary warm-up profile in accordance with the starting temperature (cold engine before cranking) and the length of time after starting. Multiple positions on the cam profile allowed for a gradual slowdown from cold-engine speed to curb idle speed during warm-up. A second feature of the cold-engine speed control was automatic kickdown from high-cam (fast-idle) engine speed to some intermediate speed. This was accomplished by the computer vacuum signal to the automatic kickdown motor, which physically moved the high-speed cam a predetermined time after the engine started.

The digital interactive control feature in the system was developed and patented by Ford.

EEC-IV

Some applications of the EEC-III were replaced for 1984 by the EEC-IV, a more versatile system with simplified componentry. It acts several times faster in

135

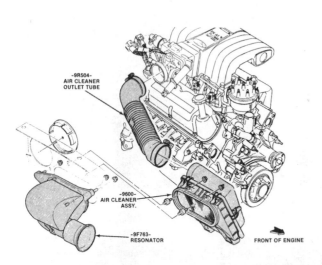

Air intake system on the multipoint injection system for the Thunderbird and Cougar five-liter V-8 engine feeds into an airflow meter mounted on top of the intake manifold.

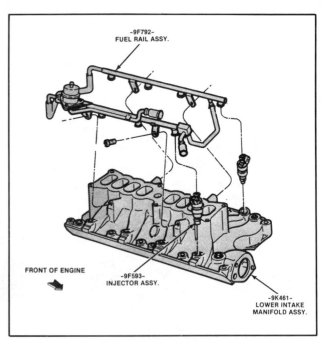

External fuel rails deliver fuel at 39 psi pressure to the port-mounted injectors in Ford's sequentially timed injection system. This is the setup for the five-liter HO Mustang V-8 engine.

its response to altered conditions, and it has twenty percent greater memory capacity despite two-thirds fewer large-scale integrated circuit chips.

According to David F. Hagen, Ford's chief engineer for engine engineering in 1984, the EEC-IV computer and memory system had as its goal the use of this management system through 1990. It was the basis of interactive spark timing and boost control for the port-injected 2.3 liter four-cylinder turbocharged engine in 1983. Here, the throttle body was eliminated, and injector nozzles were mounted on the intake manifold a short way upstream of the valves, but the electronic controls remained similar.

The injectors operated simultaneously, in pairs, once during each crankshaft revolution. The EEC-IV also had provision for more sophisticated concepts such as individual cylinder-based spark-retard control, adaptive fuel-flow control and transient fuel compensation.

By 1987, Ford had standardized EEC-IV on all made-in-the-USA engines—with low-volume exceptions such as the police model of the 5.8 liter V-8 and some truck versions of the same engine, which retained the carburetor.

For 1988, multiport injection was made standard for the new 3.8 liter V-6 engine offered in the Ford Taurus and Mercury Sable, using the EEC-IV electronic control module. Multiport injection was also extended to the 2.3 liter HSC and HSO four-cylinder engines offered in the 1988 Ford Tempo and Mercury Topaz. At the same time, multiport injection replaced single-point throttle body injection on the 5.8 liter and 7.5 liter V-8 engines offered in the Ford F-series pickup trucks, Bronco, Club Wagon and Econoline van. Single-point injection was retained for the 1.9 liter four-cylinder engine used in the Escort, though a multipoint version was also offered on this engine. The 2.5 liter four-cylinder HSC engine for the Ford Taurus also continued with single-point injection.

To develop a timed, sequential injection version of the multipoint system was mainly a matter of software reprogramming. The basic hardware already possessed the required capability.

A sequential injection multiport system was developed in combination with sonic-wave electronic exhaust gas recirculation, electronically managed Thermactor, bypass-air idle stabilization and thick-film ignition. Ready for production early in 1987, it was adopted for the five-liter V-8 engines installed in the 1988 model Ford Mustang, Ford and Mercury Town Cars, Ford Thunderbird and Mercury Cougar, and Lincoln Continental Mark VII and Mark VII LSC.

20

Pierburg's path to the Ecojet

Today's Pierburg Ecojet S is a single-point (manifold) injection system with digital electronic control. Announced in August 1987, it has not yet been adopted as original equipment by an auto maker. Pierburg is Germany's biggest carburetor manufacturer.

Formerly Deutsche Vergaser Gesellschaft (DVG), it now belongs to the Rheinmetall group.

Before the company and all its products were renamed Pierburg in honor of DVG founder Alfred Pierburg, the carburetors wore the trademarks Solex

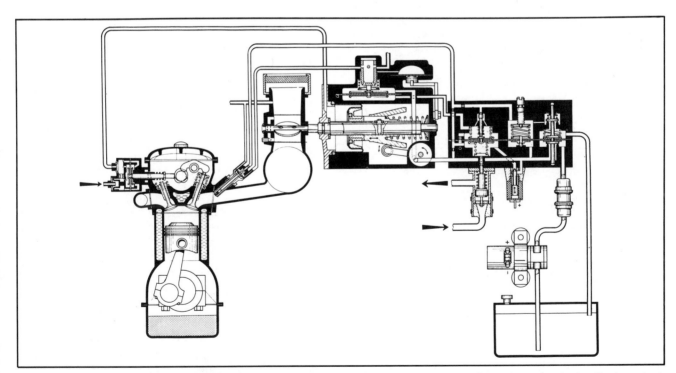

The original Zenith system did not have airflow measurement. Fuel was pumped to the pressure control unit, which was also connected to the fuel-metering unit. A cone-and-follower arrangement gave mechanical adjustments in fuel quantity.

and Zenith. And DVG's early experiments were carried out under the Zenith banner.

The first Zenith system of which details were released appeared in 1973 and was identified by the letter C. Two preceding systems, A and B, never made it through the experimental stage.

It was probably natural for a carburetor manufacturer to devise a continuous injection system that works with uncommonly low pressure, in which atomization starts within the injector nozzle. The company's vast experience with carburetors of all types has influenced its choice of mechanical linkages and hydraulic and pneumatic connections where a fuel injection expert would use electrical or electronic means.

On the Zenith C, only the fuel-feed pump was electrical. This pump drew fuel from the tank and pushed it through a filter on its way to the pressure regulator.

The pressure regulator was a valve that leveled off the line pressure to a constant value of 30 to 45 psi (according to installation), and its output line led straight to the metering-distributing unit. This unit was the heart of the whole system. It mixed fuel with air in accordance with two separate measurement devices and then delivered a uniform mixture to an individual injector for each cylinder intake port.

The translation from input measurements to output commands was centered on an eccentric cone. This cone was keyed to its spindle for rotation but was free to move axially. The cone was displaced axially on the shaft in direct proportion to the movement of a diaphragm air pump driven from the engine camshaft. The air pressure from the pump served as a measure of engine speed (rpm). This pressure was piped to a roll-sock type of diaphragm on one end of the cone spindle, displacing the cone against the load of a coil spring along its own spindle. The spindle ran horizontally.

The air pressure was applied on the small end of the cone, and the spring load was applied on the base end. A roller riding on the surface of the cone, but held in a fixed vertical arc by a pivoted lever, was therefore kept in a raised position by spring force but was allowed to descend to lower heights when the cone was moved by diaphragm force. Because of the eccentricity of the cone, its rotation also affected the position of the roller.

The cone spindle was connected to the throttle plate by means of an adjustable, mechanical coupling

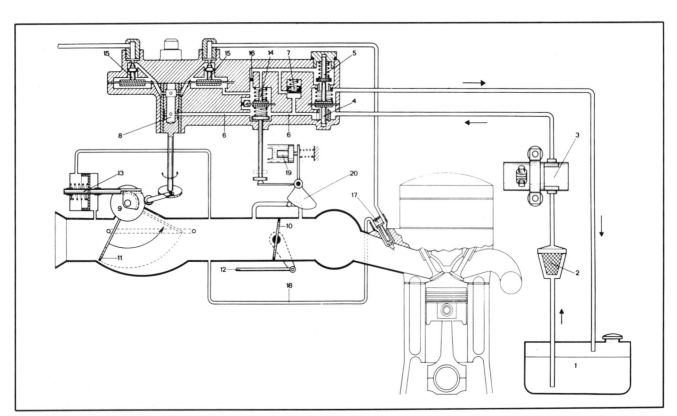

The eventual Zenith CL system embodied airflow measurement with a pivoted flap (11) positioned upstream of the throttle valve (10). 1-Tank, 2-Filter, 3-Pump, 4-Holding valve, 5-Return valve, 6-Supply line, 7-Pressure valve, 8-Metering distributor, 9-Cone, 10-Throttle valve, 11-Pivoted flap, 12-Throttle linkage, 13-Amplifier, 14-Control pressure valve, 15-Differential pressure valve, 16-Restrictor, 17-Injector, 18-Atomization air line, 19-Expander element, 20-Supplementary air valve.

that was arranged to give an indication of load (throttle opening). Opening or closing the throttle plate produced rotation in the spindle and, along with it, the keyed-on cone.

The idea of the cone may have been borrowed from Kugelfischer. It served the same purpose, namely that of having each position of the roller on the surface of the cone correspond to a specific combination of engine speed and load. Consequently, both of these input measurements would be expressed in a single-plane motion, the vertical arc described by the pivoted lever carrying the roller.

This lever conveyed the fuel requirement message to a circular disc valve mounted on the same pivot shaft. Near its periphery, this disc had an open axial channel. The disc valve housing was inserted in the fuel line. It had an inlet port and an outlet port, which registered with the channel in the disc. Disc rotation moved the channel relative to the ports so as to open or restrict the fuel flow through the valve.

From the disc (metering) valve, the fuel flow continued to the distribution section of the unit. This consisted of a base chamber with a diaphragm and an upper chamber with a number of outlet ports corresponding to the number of cylinders in the engine.

The fuel line entered the base chamber on the upper side of the diaphragm. The diaphragm opened and closed a valve on the entry to the upper chamber. Counteracting the fuel line pressure on the upper side of the diaphragm was another force: a secondary hydraulic pressure bled off from the pressure control valve.

Diaphragm (and valve) movement was transmitted to a valve block inside the upper chamber, which had an orifice for each outlet. Displacement of this valve block regulated the size of each outlet orifice. Consequently, the diaphragm worked as a differential pressure valve, whose duty it was to equalize the fuel delivery to each injector.

Any change in the load on the engine produced a transitory change in the pressure drop across the disc valve. This immediately resulted in a displacement of

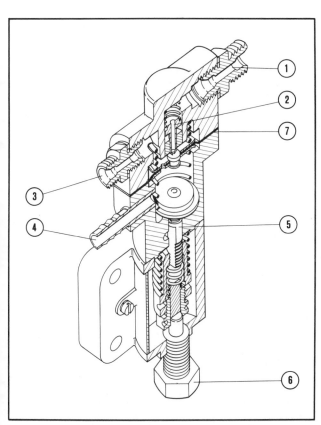

Control pressure valve of the CL system. 1-Return line connection, 2-Valve seat, 3-System pressure adjustment screw, 4-Fuel inlet line, 5-Pushrod with spring-loaded plate, 6-Adjustment screw, 7-Diaphragm.

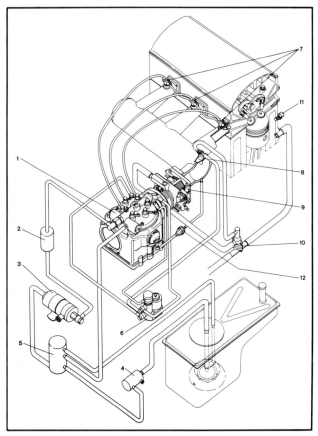

Schematic view of the DL system. Fuel is drawn from the tank by the feed pump (4) and reaches the pressure pump (3) through an intermediate tank (5). Pressurized fuel reaches the control unit (1) through a final filter (2), and is distributed to the injectors (7). A cold-starting valve (8) provides fuel enrichment as needed, and a supplementary air valve (10) ensures a stable idle. Pressure in the hydraulic control system is maintained by a pressure valve (6). A valve (12) gives protection from backfiring. A thermo-time switch (11) limits warm-up enrichment. The throttle body is 9.

the distributor diaphragm, so that a constant pressure drop was reestablished.

The system included a form of acceleration pump. This pump worked by means of an offset cam on the cone spindle. The cam responded to opening of the throttle by causing a sudden increase in the fuel pressure inside the chamber above the distributor diaphragm. That kept the valve open so as to admit more fuel into the distributor valve.

The injector nozzle was positioned closely behind the inlet valve head, mounted in a mantle and separated from the nozzle body by an open space over most of its length. The nozzle injected a very fine, low-pressure spray, which began to atomize inside the injector mantle (an idea that was later embraced by Bosch).

An open connection between the nozzle and the mouth of the throttle venturi supplied atomization air into the space inside the mantle. This connection had its greatest effect during conditions of high pressure differentials, such as at idle and in part-load operation. The injector valves were closed when the engine was not running. Once the engine was started, the fuel pressure had to overcome the spring load on the needle and a modulated resistance pressure provided by a hydraulic line from the pressure regulator.

The system also included a cold-start device that consisted of a magnetic valve governed by a thermo-time switch. It supplied additional fuel into the line behind the disc valve.

A warm-up valve ensured the gradual lessening of mixture enrichment during warm-up. The closed space below the diaphragm, in the lower chamber of the distribution section, was pressurized under normal operating temperatures. This pressure acted to close the warm-up valve in response to a thermostat submerged in the coolant.

Closure of the warm-up valve reduced the pressure on the lower face of the diaphragm so as to admit more fuel until the engine reached normal operating temperature. Simultaneously, it turned the throttle plate so as to add sufficient air. This gave a fast-idle setting for as long as necessary.

A no-return valve on the entry port to the pressure regulator made sure that the system remained pressurized when the ignition was turned off. This prevented evaporation and vapor lock and facilitated hot-starts.

The Zenith C gave excellent results under wide-open-throttle acceleration, but it was not so satisfac-

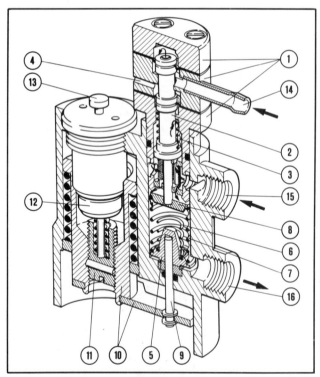

Control-pressure-regulating valve for the first Zenith single-point injection system. This valve controls the pressure that balances the fuel delivery pressure across the diaphragm in the differential pressure valve. 1-Valve body, 2-Pushrod with diaphragm, 3-Spring, 4-Vent hole, 5-Spring base, 6 and 7-Differential pressure springs, 8-Valve plate, 9-Pressure rod, 10-Stroke transfer link, 11-Adjustment screw, 12-Heat-sensitive element, 13-Electrical terminal, 14-Manifold vacuum connection, 15-System pressure in, 16-System pressure out.

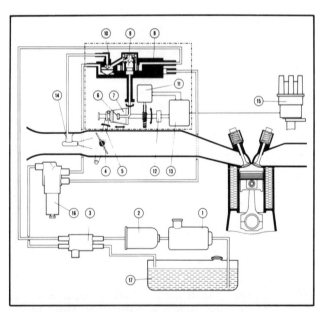

Initial Zenith single-point injection system. One single injector is aimed into the throttle opening and is metered by electronic control. 1-Electric fuel pump, 2-Fuel filter, 3-Pressure-holding valve, 4-Throttle plate, 5-Axial movement of the cone, 6-Cone follower, 7-Cone, 8-Metering distributor, 9-Metering valve, 10-Differential pressure valve, 11-Servomotor, 12-Potentiometer, 13-Electronic control unit, 14-Injector, 15-Ignition distributor, 16-Control pressure regulator valve, 17-Tank.

tory in a variety of transient conditions met in everyday driving. Consequently, it tended to be used only in racing cars.

In an attempt to overcome these objections, Pierburg introduced the Zenith CL (the *L* stands for *Luft*— German for "air"—to indicate that this system used airflow metering). Here again, Pierburg did not copy Bosch but went out inventing on its own. It was not long after the introduction of the Zenith CL system that Bosch purchased a big block of shares in Pierburg.

In the Zenith CL, Pierburg chose to put a flap in the intake duct, upstream of the throttle plate, to monitor the mass airflow. The flap angle was determined by the pressure differential between the open area nearest the inlet and the partial depression created in the section between the flap and the throttle plate. This pressure differential was measured across a small hole in the flap. The flap was L-shaped in profile and pivoted on a spindle located below the duct. It was spring-loaded toward the closed position, where a bulge on the duct provided a simple form of bypass for idle air and so on.

An accordion bellows on the inboard side of the flap acted as a damper on flap movement during airflow fluctuations, inertia reactions and so on. The movements of a three-dimensional cam, mounted on the flap shaft, were picked up by a pivoted lever with a roller at its free end. This mechanism replaced the cone used in the Zenith C system. The three-dimensional cam reacted to changes in mass airflow by moving radially, pivoting on its shaft.

There was no longer any camshaft-driven air pump to give rpm readings by way of a diaphragm. However, there was a vacuum connection from the manifold, downstream of the throttle, direct to a diaphragm carried on the shaft of the roller running on the cam track. Movement of this diaphragm would displace the roller axially so as to move it to a different part of the cam surface.

The pivot arm that carried the roller transmitted its swinging motion directly to the piston in the metering unit. This was a new design based on the same principle of sliding ports.

The basic principle used for the metering and distribution depended on control of the average pressure in the distributor unit, which consisted of a piston, a sleeve and a ring. The piston was turned by the pivot arm while the sleeve was held still. This control piston had a number of triangular slots, one for each cylinder in the engine. The sleeve surrounding the piston had similar slots, matching in size and number. Rotation of the piston inside the sleeve altered the cross section of the slots that were open to let the fuel pass, and the cross-sectional area served to regulate the amount of fuel delivered to the injectors.

In 1976, this system evolved into the Zenith DL. It received a separate starting valve that sprayed fuel into a spherical bulge on the manifold, downstream from the throttle, whenever a thermostatically governed pressure valve relented enough to let its pressure drop below fuel line pressure, which would open the line.

The Ecojet S components: throttle body assembly in the foreground (right) and temperature sensors (left). The electronic control unit is in the background.

Electronic control

In 1977, Pierburg introduced an electronically controlled continuous-flow port injection system for racing cars.

An electrically driven pump drew fuel from the tank and pressurized it to about 60 psi. After filtration, the fuel arrived at the metering unit and was distributed to the individual injectors. Excess fuel was spilled off and returned through a system pressure valve, a pressure-regulating valve and a depressurizing valve to the tank. The metering unit contained a spill plunger that was linked to the main throttle and an electric motor actuated from the electronic control unit.

Rpm signals reached the control unit from the ignition distributor. The control unit also received information on load from the throttle linkage.

The injectors had pintle-type nozzles. They received a small amount of bypass air from the upstream side of the throttle to ensure proper atomization during light-load operation.

Despite continual improvements, Pierburg could not convince the auto industry to buy either the CL or the DL as original equipment or even as a regular production option. During the years it developed the CL and DL, Pierburg also toyed with electronic carburetors, and it began development work on a single-point injection system to compete against the Bosch Mono-Jetronic. Eventually, all Pierburg's efforts were

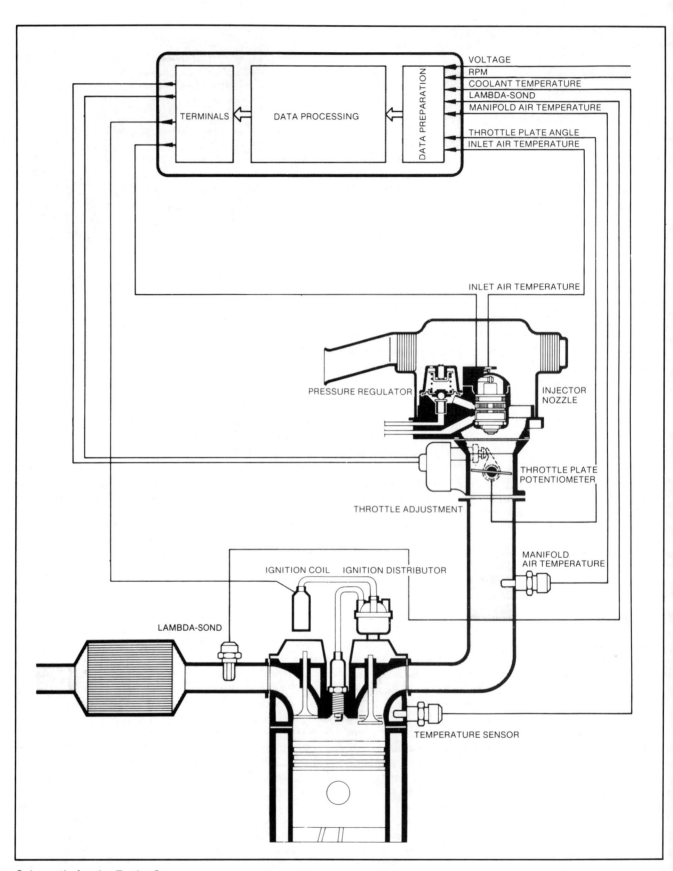

VOLTAGE
RPM
COOLANT TEMPERATURE
LAMBDA-SOND
MANIFOLD AIR TEMPERATURE

THROTTLE PLATE ANGLE
INLET AIR TEMPERATURE

TERMINALS

DATA PROCESSING

DATA PREPARATION

INLET AIR TEMPERATURE

PRESSURE REGULATOR

INJECTOR NOZZLE

THROTTLE PLATE POTENTIOMETER

THROTTLE ADJUSTMENT

MANIFOLD AIR TEMPERATURE

IGNITION COIL   IGNITION DISTRIBUTOR

LAMBDA-SOND

TEMPERATURE SENSOR

Schematic for the Ecojet S.

directed toward the low-cost single-point injection system.

Ecojet S

As early as 1980, Pierburg approached the European car makers with its EL fuel injection system. From the outside, this looked confusingly like a carburetor. Indeed, it had much in common with the air valve carburetor, including an oil-damped piston that adjusted its position in a cylinder according to manifold vacuum. The fuel metering was arranged by the same type of devices that were used in the Zenith DL, though the nozzle had different dimensions and included a passage for atomization air to get inside the mantle.

The *S* in Ecojet S stands for "single point." The system is built around the concept of central mixture formation with an indirect indication of the airflow. The digital control unit is programmed for two main criteria: the angle of the throttle plate and the rpm (with a stoichiometric air-fuel ratio).

On the basis of input received, the control unit determines not only the fuel quantity but also the spark timing. The input includes coolant temperature, manifold temperature, ambient temperature, manifold vacuum, engine speed (rpm), throttle plate angle (load) and a Lambda-Sond signal.

The ignition coil is wired to the electronic control unit, which selects the spark timing on the basis of load versus speed, modified by low temperature and/or high altitude.

The components are to a large extent inherited from the EL system. The fuel flow starts with a feed pump and filter, and goes on to a mixing unit with injector nozzle, pressure regulator, throttle valve, potentiometer, Lambda-Sond, pressure sensor and the electronic control unit with its eight-bit microprocessor.

Ecojet S works with intermittent—not continuous—injection. The pressure regulator maintains a constant pressure throughout the system. The injector valve opens in accordance with speed-based pulsing as directed by the control unit, and the fuel quantity per firing stroke is varied by altering the duration of the injector valve opening. Fuel enrichment for cold-starts is determined by data from the three temperature sensors, modified by the engine speed requirement for normal operation during the warm-up phase. Overenrichment after stalling is prevented by a thermo-time switch.

Fast-idle speed is determined by a solenoid wired to the control unit. This solenoid sets the throttle plate at a high enough angle to admit a sufficient volume of air at a particular temperature. After warm-up, idle speed is stabilized on the basis of temperature and is automatically compensated for parasitic external loads (air conditioning, automatic transmission drag and so on). Enrichment for acceleration is subject to intricate calculations that take into account engine speed and temperature as well as the throttle opening.

Full-load enrichment is provided in accordance with throttle plate angle and Lambda-Sond input. The latter tends to minimize fuel waste as well as keep emission levels down. To restrict emissions on sudden closing of the throttle, a dashpot is fitted on the throttle linkage. The dashpot slows down the closing rate in accordance with the rpm at the moment of accelerator pedal release and the throttle plate angle up to that moment.

Automatic lean-out of the mixture on a trailing throttle is ensured, as is fuel shutoff during coasting or engine braking. Smooth transitions are ensured by time-detent, which imposes a delay based on engine speed, rpm gradients (rise or fall rate) and engine (coolant and manifold) temperature.

21

Fiat/Marelli/Weber

As Europe's biggest carburetor manufacturer, Weber has been watching the advance of fuel injection for many years. Weber's pulse beat faster when Ferrari began testing direct fuel injection on its 1500 cc V-6 Formula One engines in 1961—and came up with new carburetor systems that saved the day. But the challenge would not go away, and in 1969 Weber was developing a port-type mechanical fuel injection system for the three-liter Ferrari racing engine.

The Weber company in Bologna came under Fiat ownership in 1951. Fiat also controls Magneti Marelli, makers of ignition systems and other electrical equipment. Marelli produced the electronic ignition system that was standard on the Fiat Dino in 1967-69.

In association with Magneti Marelli and Fiat's research laboratories, Weber began experiments with electronic fuel injection in 1972, using analog computer technology.

Details of an experimental system were revealed in 1977. It used an ultrasonic airflow meter invented by R. Rinolfi of Fiat's research center and a digital microprocessor developed by Marelli's electronics wizards.

Fuel metering was calculated on the basis of signals from an engine speed sensor, engine stroke (phasing) sensor, pressure transducer, coolant temperature transducer and the ultrasonic airflow meter, which was the only truly unusual component in the system.

The flow meter was developed with the aim of being able to measure the air mass admitted into the engine at each stroke (one stroke equals one-half revolution). Its body was made up as a duct with a circular section. Two piezoelectrical microphones were mounted at fixed points along its axis. The entry duct to the flow meter was equipped with guide vanes to maintain laminar flow through the meter duct in order to obtain maximum accuracy. The microphones were mounted to aerodynamically shaped supports developed to cause minimal disturbance in the flow field during measurement.

When excited by a step in the voltage supplied from the control circuit, the microphones emitted ultrasonic energy. The maximum amplitude of the ultrasonic wave set up this way corresponded to the natural frequency of the microphones, which was about 300,000 hertz (cycles per second). The moment the ultrasonic energy emission stopped, the microphones began to work as receivers for each other's waves.

The two ultrasound waves were emitted simultaneously, every millisecond. The relative delay between the moments of reception at opposite microphones—or, in other words, the difference in elapsed time between the travel of the two waves—provided information from which the air velocity could be accurately calculated. By comparing the velocity signal with pressure and temperature readings, the wave could be made to express a measure of mass airflow.

Downstream from the meter duct was an acoustic filter, which had the task of protecting the high-frequency components from sound perturbations emanating from the engine. Flow meter tests proved that the unit was able to measure accurately (within a one percent margin of error) flow masses throughout a range from 0.02 to 0.4 pounds per second with velocities varying between 4.5 and 165 feet per second.

The error was attributed to boundary layer conditions through the metered section of the duct. Ambient air temperature did not affect the flow pattern or the wave formation.

The control circuit picked up the time signals, which were first amplified, then filtered and finally fed to logic-comparing circuits, from which they emerged transformed into a square electronic wave of specific length, proportional to mass airflow.

In a second stage, this wave was processed further and compared with inputs from other trans-

ducers. The resulting output indicated air mass inflow during a single engine stroke. The requisite comparison data were supplied from the engine speed and stroke sensors.

The engine speed sensor was an induction coil that sensed the impulse frequency of magnetic pins attached to the flywheel. The stroke sensor was an analog device that gave off a pulse every time it detected the passing of a reference mark on the camshaft.

This mark signaled that the piston in cylinder number one was at top dead center, which gave the control unit a fix on the piston positions in all the

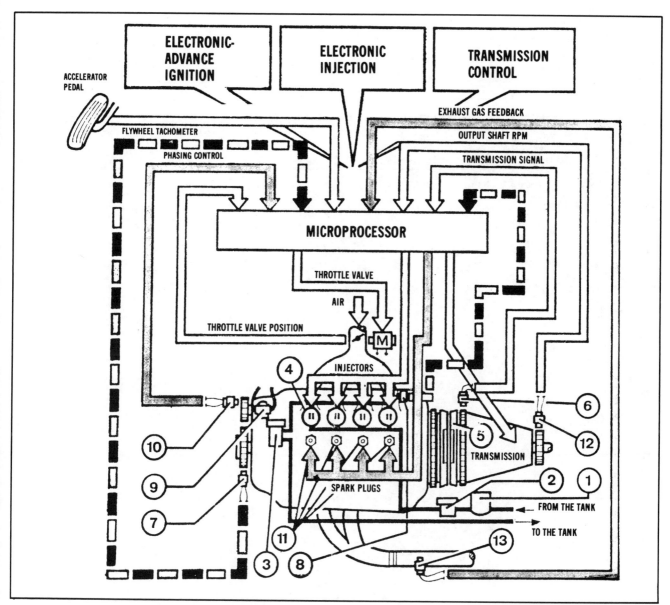

In the Marelli control unit created in 1973, control programs included the automatic transmission as well as the spark timing.

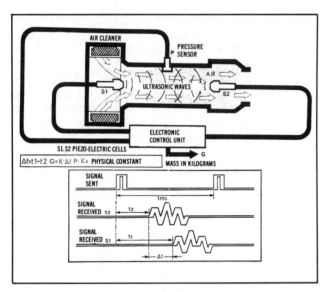

Within the schematic:

AIR CLEANER

PRESSURE P SENSOR

AIR

S1

ULTRASONIC WAVES

S2

ELECTRONIC CONTROL UNIT

S1.S2 PIEZO-ELECTRIC CELLS

$\Delta t = t1 - t2$ $G = K \cdot \Delta t \cdot P$ $K =$ PHYSICAL CONSTANT

G

MASS IN KILOGRAMS

SIGNAL SENT

SIGNAL RECEIVED S2 — t2

SIGNAL RECEIVED S1 — t1

1ms

Δt

Schematic of Marelli's ultrasonic flow meter.

other cylinders (in taking account of rpm signals). Signals from the rpm sensor and stroke sensor were fed to a preprocessing block that converted them into basic logic data for synchronizing the complete system.

All the input data were translated into binary code before entering special registers where the microprocessor could read them.

The microprocessor was a digital computer capable of handling eight-bit words. It had two separate memories, a programmable read-only memory (PROM) and a random-access memory (RAM).

The PROM contained the control program instructions, and the RAM stored the input variables indicating the instantaneous engine operating conditions. The control module output signals determined both the spark advance and the opening duration of the injector valves.

Fuel was taken from the tank by an electrical feed pump, pushed through a filter and fed to the pressure-

Experimental installation from 1983 on a Fiat Argenta, with the 1608 cc four-cylinder twin-cam engine. The ultrasonic airflow meter had been abandoned in favor of Weber's manifold vacuum measuring device.

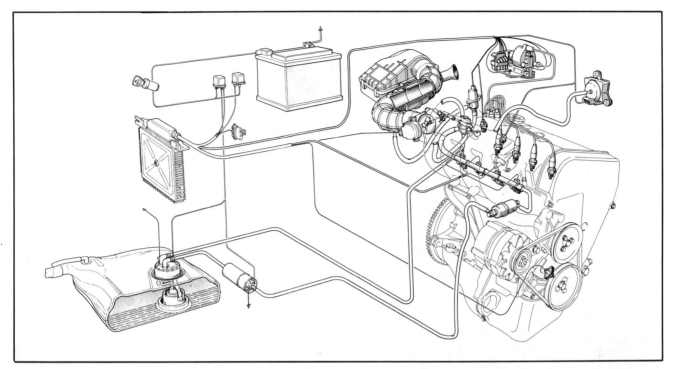

The Weber fuel injection system installed on the Lancia Delta GT i.e. In 1986 it gave an idea of how the system could be applied to a mass-production engine without breaking off with any of the basic principles of the concept.

regulating valve. Air entered the ultrasonic flow meter straight from the air cleaner and flowed from there to the throttle valve. Injectors were mounted at the intake ports. These had timed delivery through electromagnetic valves as directed from the control unit.

Fiat's experiments with this system, spread out over an eighteen-month period in 1978-79, were aimed principally at reducing emission levels by maintaining a strictly constant air-fuel ratio (assisted by a Lambda-Sond oxygen sensor). The experience was valuable: After an initial analysis of the results, Weber began a definition program for a production model injection system in 1980.

That blossomed into a full-scale development program led by Weber's director of advanced engineering, Valerio Bianchi. The program had a generous budget and a staff that began with thirty-five engineers and technicians and gradually increased to eighty people.

At the concept stage, cost was just one of seven criteria to be considered and weighed against six others such as emissions, fuel economy, power and torque, reliability, drivability and the ability to operate when defective. Weber wanted superiority in a maximum of areas, and the analysis pointed straight to one goal: timed, sequential, multipoint injection including ignition control correlation between fuel metering and spark timing, so as to constitute a complete engine

management system and not just something to replace the carburetor.

During the development phase, the Weber engineers discovered that the biggest difficulties lay not in

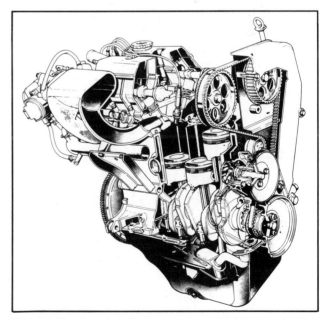

The 1.6 liter Lancia twin-cam engine with Weber injection shows the port-mounted injector nozzles, supplied with fuel from a pressurized rail.

the electronics hardware, but in finding reliable, accurate and fast-reporting sensors. Another aspect of the development that demanded considerable time was the development of an electromagnetic injector that was capable of opening and closing in less than a millisecond.

Gradually, a more-than-satisfactory system took

The Marelli/Weber system coordinates fuel injection and spark timing. Each cylinder has two inlet ports, all fitted with injection nozzles (16 in this V-8).

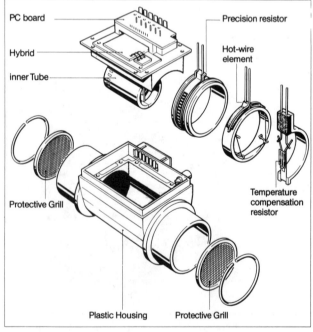

Exploded view of the Bosch hot-wire airflow sensor, revealing the sheer simplicity of its construction.

shape. After a baptism by fire on the Lancia Group C racing cars in 1983, it was adopted for the new Ferrari Turbo GTO in 1984.

Weber calls its system "phased sequential" because the microprocessor puts the sequence under separate control from the metering. The sequence control depends on the number of cylinders, while the metering control is linked to the system's capacity for variations in timing.

The base injection period is calculated from information about manifold absolute pressure, air temperature in the manifold and engine speed. It is modified, according to need, by inputs from other sensors, such as coolant temperature during warm-up, acceleration (throttle position and the manifold pressure derivative), deceleration (throttle position and engine speed) and starting (elapsed time since switchoff and coolant temperature).

With speed and accuracy as the main goals, Weber's search for a superior type of airflow meter never ceased. Entertaining some doubts about certain aspects of the scientifically interesting ultrasonic airflow meter developed by Fiat and Marelli, Weber explored simpler devices, using speed-density measurements to get a precise reading on manifold absolute pressure. Such a device was part of the system used on the Lancia racing cars in 1983.

On the production-type system, the airflow meter is inserted at the manifold end of the fresh air duct from the air cleaner. Inside the sensor, the piezo-resistive effect of elements deposited by the thick-film technique ensures rapid information about pressure changes.

How fast is this system? Detecting a change in operating conditions, feeding fresh input to the microprocessor, processing the data from all sensors and issuing new orders is accomplished in one engine cycle. In a four-cylinder engine, that means in half a revolution. If the engine is doing 3000 rpm, that corresponds to 0.1 millisecond.

By this method of manifold pressure measurement, Weber also obtains data on the engine's volumetric efficiency. These data are stored in the microprocessor memory as revolution and load functions, and thereby serve to fine-tune the fuel metering. And Weber's system is fully compatible with additional functions such as automatic idle speed stabilization and closed-loop air-fuel ratio adjustment.

How does the Weber system compare with the Bosch Motronic? The two systems have emerged from different backgrounds. Bosch had years of electronic injection and ignition experience before combining the two, while Weber is a newcomer. The existence of a specific industrial fabric in the Bosch plants may have influenced the composition of the Motronic system; for Weber and Marelli, all manufacturing equipment had to be ordered new.

Finally, there is the difference in mentality. Motronic came into being through close cooperation with BMW, which was never shy about keeping its suppliers aware of its price targets. In contrast, Weber was able to select the theoretically ideal principles in relative freedom from cost considerations.

Despite having digital control on the Motronic, which is easily adaptable to sequential injection, Bosch arranged the injector valves in parallel, so that they open and close all at the same time, without regard for the firing order. Why? This enabled Bosch to miniaturize the electronic switching.

As a result, there are cylinder-to-cylinder variations in the residence time of the fuel injected into each port. To counteract this, the injectors are set to deliver the metered amount in two portions: half the amount twice during each camshaft revolution. This way, Bosch avoids any fixed relationship between the cam angle and the moment of injection, which allows triggering of the injection by the ignition contact breaker.

How important is this in practice? Does it make any difference in fuel economy? According to Weber, the exact advantage it has over Bosch depends on car-to-car variations. However, in the fuel consumption test cycles, the Weber system undercuts Motronic

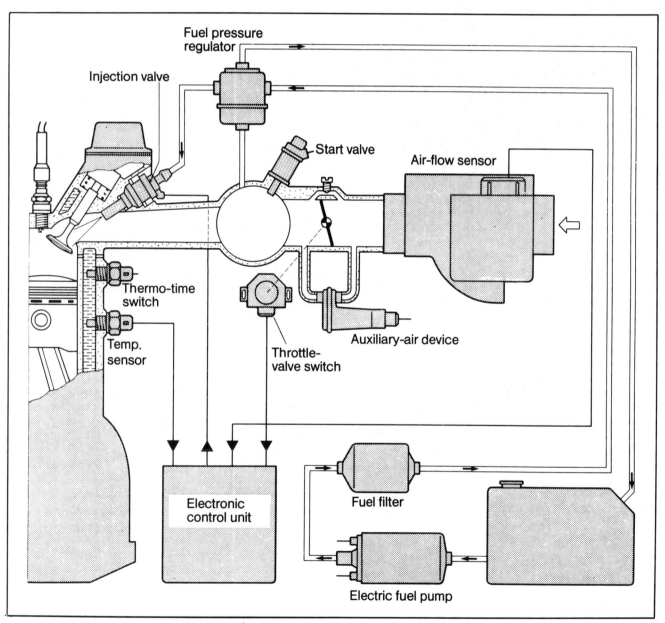

Schematic for the LH-Jetronic system including a Lambda-Sond oxygen sensor. The hot-wire airflow sensor replaced the flap valve.

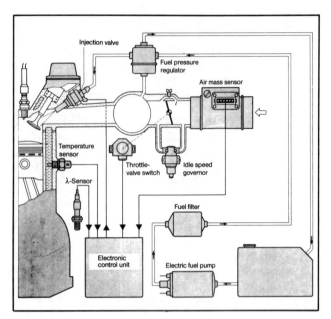

Schematic of the LE-Jetronic system, with its flap-type airflow sensor. A thermo-time switch was added in the water jacket to adjust warmup enrichment.

by margins from 0.2 to 0.7 liter per 100 km (0.007 to 0.025 pounds per mile) at speeds of 90 and 120 kph (56 and 74.5 mph). This means an average difference of 2.3 miles per gallon (33.6 rather than 31.3) in Weber's favor.

Lancia picked Weber fuel injection for the Delta S.4 rally car that won the 1985 world championship. The Delta S.4 had a four-cylinder 1.8 liter twin-cam engine that put out over 400 hp (under the impulsion of two-stage supercharging [Roots blower plus turbo] and air-to-air intercooling).

For 1986, Weber fuel injection became standard equipment on the Lancia Delta GT i.e., powered by a 1.6 liter version of the same twin-cam design, rated at 108 hp. And for 1988, a turbocharged two-liter version with intercooling, rated at 185 hp, was used in the Lancia Delta HF 4WD (four-wheel drive).

Ford adopted Weber fuel injection for its 1986 model Sierra RS Cosworth, powered by a 204 hp turbocharged sixteen-valve four-cylinder two-liter engine. Aston Martin picked Weber's system for its new Zagato-bodied coupe, unveiled in March 1986, which was powered by the 5.2 liter four-camshaft V-8 and was rated at 432 hp. The same basic engine fitted with four two-barrel downdraft carburetors put out 306 hp. The Weber-injected engine has been extended to the 1986 V-8 Saloon, Volante convertible and Lagonda four-door limousine.

For the Ferrari F-40, first shown in September 1987, Weber contributed an injection system with additional refinements. The three-liter V-8 engine has four valves per cylinder; consequently, Weber put an injector nozzle in each inlet port (sixteen in all).

Turbocharge boost is coordinated with the fuel metering and spark timing by means of a bypass valve (waste gate), and turbocharge pressure signals fed to the electronic control unit are used in the fuel-metering calculations. With an internal compression ratio of 7.8:1 and twin IHI (Ishikawajima-Harima Heavy Industries) turbochargers plus air-to-air intercooling, this unit delivers 478 hp at 7000 rpm, with a maximum torque of 577 Nm at 4000 rpm. Top speed is about 320 kph (over 200 mph) and acceleration from zero to 200 kph (124.3 mph) takes only twelve seconds.

22

Solex single-point injection

Since July 1986, Solex has been allied with Weber, both members of a new industrial group jointly owned by Fiat (sixty-five percent) and Matra (thirty-five percent). The group also includes Veglia-Borletti and Jaeger (instruments). Before July 1986, Solex and Weber were working independently on both single-point and multipoint fuel injection systems. Since then, Solex has concentrated on the single-point system, while Weber has decided not to go further in that direction.

Weber revealed its first single-point injection system to the technical press in November 1984. It was a very simple concept, combined with the electronic control unit for the ignition system. The two basic inputs for calculating the amount of fuel to be injected were engine speed (rpm) and throttle plate position (load). The metering was corrected by data on ambient air and coolant temperature.

Weber's system was never used on production cars in its original form, but it was adopted by Fiat in a low-cost version developed by Fiat's American liaison

The components of the Solex single-point injection system as they looked in 1986. Miniaturization may be possible before it goes into production.

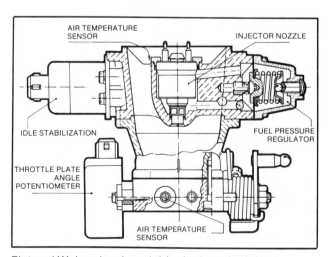

Fiat and Weber developed this single-point injection system for the two-liter twin-cam engine used in the 1987 Regata 100-S.

office in Detroit. General Motors agreed to cooperate, and the result was a system based on a Weber throttle body with a Rochester injector nozzle. The control unit was made by Delco Electronics, the sensors by Borletti and the fuel pump by AC Delco. This was combined with Marelli electronic ignition and fitted to the Fiat Regata 100-S as standard equipment, beginning in June 1986.

Solex displayed its first-generation single-point injection system at the Equip-Auto exhibition in Paris in 1985. The company hoped to attract business based on the new and stricter emission control standards in several European markets (Sweden, Switzerland and Germany encouraged the use of lead-free gasoline and catalytic converters; France and Italy were opposed to the standards and the devices).

In terms of operating principles and hardware, Solex did not stray far from the broad lines traced by Bendix, but the Solex people did set out to develop new and original software for control of its injection system.

The Solex system was built up around a throttle body with a single injector nozzle. The throttle body also contained the throttle plate angle potentiometer, fuel pressure regulator and supplementary air bypass.

A low-pressure (10 psi) fuel pump located inside the tank ensured a continuous gasoline supply. The electronic control unit received input signals concerning engine speed, ambient air and coolant temperature, and fuel pressure. It also had provision for closed-loop feedback (Lambda-Sond) data on oxygen content in the exhaust gas. Output signals metered the amount of fuel to be injected and timed the spark according to conditions.

Solex even developed its own software for controlling idle air volume (idle stabilization) and specific antipollution functions such as engine rpm deceleration rate control when the accelerator is suddenly released (dashpot). It also incorporated fuel circuits with evaporation control canisters.

At the Paris auto show in October 1986, Solex indicated a probable delivery date "at the end of 1988" for a production system, but experimental sets were supplied to the auto industry before the end of 1987.

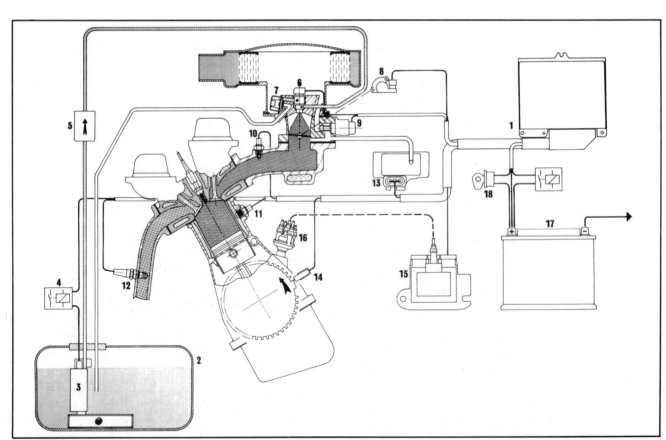

Schematic of the Solex single-point injection system. 1-Electronic control unit, 2-Fuel tank, 3-Fuel pump, 4-Relay, 5-Fuel filter, 6-Injector nozzle, 7-Fuel pressure regulator, 8-Throttle plate angle potentiometer, 9-Supplementary air bypass, 10-Mixture temperature sensor, 11-Coolant temperature sensor, 12-Lambda-Sond, 13-Manifold vacuum sensor, 14-Crank angle and rpm sensor, 15-Ignition module with coil, 16-Ignition distributor, 17-Battery, 18-Ignition and starter key.

23

The future of fuel mixture management

This chapter was prepared under the working title "Where do we go from here?" And that's an impossible question to answer.

The strides made in the past twelve to fifteen years have been so enormous that we have now arrived at a platform where a variety of advanced

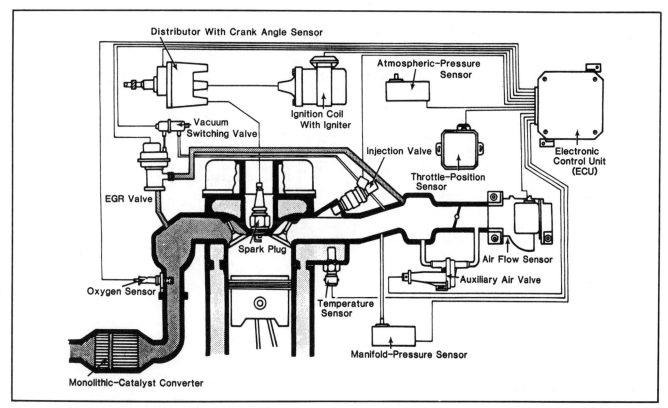

Toyota experimental engine management system from 1980, built around the Nippondenso (Bosch) L-Jetronic.

This system included spark timing and exhaust gas recirculation.

E.C.C.S. DIAGRAM—CA18ET Engine

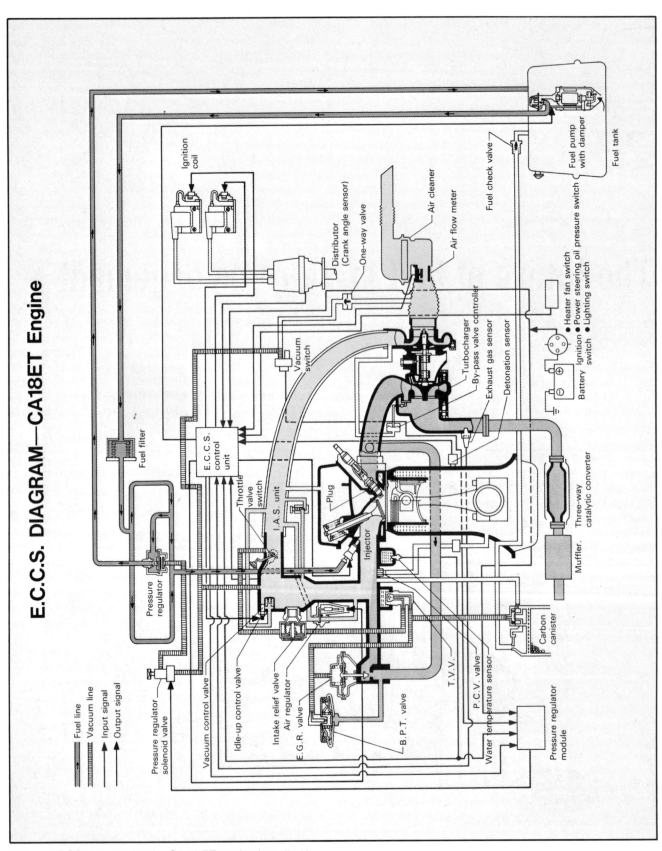

Nissan ECCS diagram for the CA 18 ET engine installed in
the 1981 Cedric and Gloria.

154

technologies are available, but their possible links to the fuel systems exist only in a mist of technological surfeit. We need more sharply defined goals than at present before we can get a reliable compass reading on the direction that can and will bring tangible advantages.

Our view of the future is limited to a study of the ideas that are still being kicked around in the laboratories, waiting for their time to come, which could lead us to certain tentative conclusions with regard to technical trends.

Manufacturing and marketing trends are subject to different laws entirely. They are more dependent on economic and political events than on a breakthrough in construction materials, fuel chemistry, electronics, robotics or pure physics.

For example, the price a car owner is willing to pay for a more efficient fuel system is going to depend on fuel supply (which dictates crude oil pricing) and the cost of alternative fuels, which may require big differences in the mixture preparation methodology and equipment.

In addition, legislation is and will likely remain a strong influence on anything that will go into our future cars. Resulting changes can be radical; witness, for instance, the switch to gasohol in Brazil, with a concurrent reservation of diesel fuel for commercial vehicles.

The following discussion ignores such factors and concentrates on the technological possibilities.

Complexity under the rising sun

Toyota had been experimenting with lean-burn engines since about 1972, but it was not until 1984 that the existence was revealed of a fuel-injected version of the 4A-ELU engine of 1587 cc displacement. This engine was fitted with a three-way catalyst and closed-loop feedback, which could save up to twenty-two percent of the gasoline burned by the standard version of the same engine (in the same vehicle, in the US test cycles). The control logic was arranged so as to provide an air/fuel ratio of 15.5:1 at idle and 16:1 at maximum rpm. At constant speed, the mixture was leaned out to 21.5:1.

In 1979-80, Toyota experimented with new concepts in software so as to obtain asynchronous control logic for D-Jetronic and L-Jetronic systems fitted on the two-liter six-cylinder engine. The underlying idea was to use compensating and adaptive circuits in conjunction with spark timing to minimize torque fluctuations.

In normal operation, all injector valves were operated simultaneously at the start of the intake stroke in cylinders one and six. In addition, all injector valves could be operated asynchronously by an acceleration signal. Fuel injection pulse width was determined by engine speed and manifold pressure (D-Jetronic) or mass airflow (L-Jetronic) modified by warm-up, transient and feedback conditions.

EGR and idle speed control were included in the system. A pneumatic EGR valve was controlled by an electric vacuum switching valve as a function of manifold pressure and rpm. Idle speed was controlled by an electrically operated auxiliary air valve whose task was to modulate the amount of bypass air in proportion to the pulse width dictated by the control unit. It not only served to maintain a steady idle speed during warm-up, but also acted as a dashpot and fast-idle device.

In its original form, the system was not effective in controlling torque fluctuations and gave a poor conversion factor in the three-way catalyst. A modified system in which all injectors are operated instantaneously when the throttle valve is opened cured these problems well enough to satisfy the investigators.

Nissan announced its ECCS (electronic concentrated engine control system) in 1980. As applied to the 2.8 liter six-cylinder engine, it gave a seven to ten percent drop in fuel consumption while satisfying contemporary Japanese, California and United States (forty-nine-state) emission control standards.

Basically, the ECCS was a microprocessor control system for managing the air/fuel ratio, EGR, spark timing and idle speed. Using the L-Jetronic as a base, the Nissan engineers added a vacuum modulator valve to control the EGR and auxiliary air valve. Fuel was injected once per crankshaft revolution, with acceleration enrichment ensured by an additional injection. The spark-advance mechanism was eliminated from

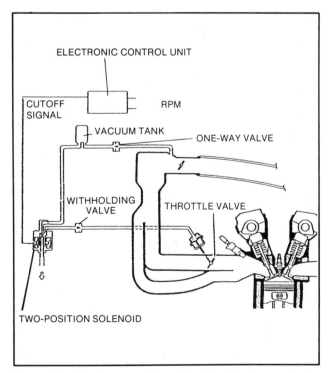

Staged fuel feed was used on the 16-valve four-cylinder engine in the 1987 Nissan Sunny.

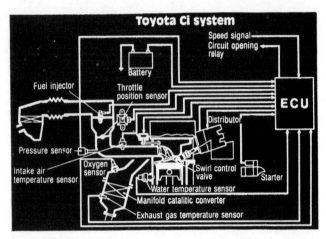

Toyota's central injection system from 1985 was developed for variable lean-burn combustion.

the distributor and the ignition coil got its base current from the control unit.

The logic was arranged so that when NO_x (oxides of nitrogen) emissions remained within tolerable levels, the spark timing and EGR rate were optimized for the best fuel economy. In normal driving, EGR provided the principal means of NO_x control, but in high-speed operation, under heavy load and so on, the NO_x emissions were restricted by playing with the spark timing and air-fuel ratio, for optimum catalyst efficiency.

Toyota revealed its electronic CI (Central Injection) single-point injection system in the fall of 1985. To ensure variable lean-burn operation, it used high-pressure (36 psi) injection and helical intake ports with swirl-control valves.

The reason behind the choice of a high-pressure system was the desire to obtain proper fuel atomization in a minimum of time. The injector was located above the throttle plate, offset from the throat axis. It had a nozzle that provided a narrow fuel spray angle intended to minimize wall wetting.

The Toyota engineers claimed that their system pushed the lean-misfire limit further away. During normal acceleration, the system ran with a stoichiometric mixture (14.7 parts of air to one part gasoline). For wide-open-throttle acceleration and top-speed operation, the mixture was even enriched to 12 to 13:1. But during cruising, deceleration and idling, the system provided a lean mixture with air-fuel ratios between 16 and 24:1.

The control units received input data on throttle opening, intake manifold pressure, rpm and vehicle speed. Compared with the feedback carburetor system, the variable lean-burn engine gave ten percent better fuel economy, as installed on the 1.8 liter four-cylinder engine.

A staged air intake system was introduced on the 1598 cc sixteen-valve four-cylinder CA 16 DE engine

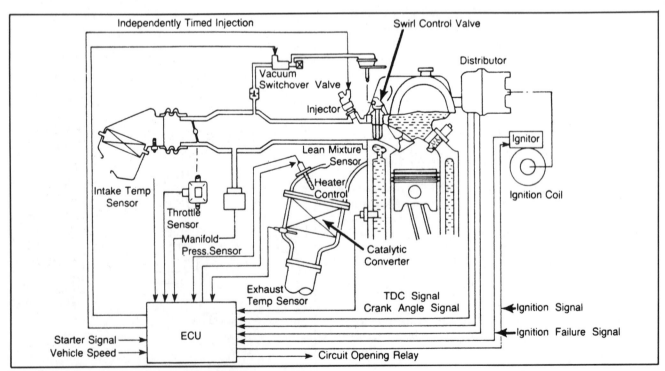

Toyota lean-burn injection system applied to the 1984 1.6 liter 4A-ELU engine (with closed-loop feedback).

of the 1987 Nissan Sunny GTi 16 S coupe. Its purpose was to improve low-end torque without altering the engine's strong top-end power output characteristics.

The intake manifold was divided into primary and secondary branches. The secondary branches were fitted with butterfly valves controlled by a solenoid that was sensitive to manifold vacuum but acted on instructions from the electronic control unit.

The butterfly valves were mounted a short distance upstream of the fuel injector nozzles. They were closed at low engine speed (below 3800 to 4000 rpm) to promote high turbulence in the primary branches and to generate conditions favorable to better fuel atomization.

The engine had a maximum output of 122 hp at 6000 rpm, and a peak torque of 138.3 Nm at 5200 rpm. The car weighed 1,080 kg and had a top speed of 205 kph. It was capable of accelerating from 0 to 100 kph in 9.7 seconds, and it had a steady-state fuel consumption of 8.6 liters per 100 km (27.3 miles per gallon) at 120 kph.

In 1985, Toyota developed a two-hole injector for multipoint injection systems. This was used in engines with four valves per cylinder. With such engines, gas velocity in the port areas can fall so low that the fuel does not mix properly with the air. Early injection is no solution because it can lead to a rise in emissions, and moving the injector valves upstream away from the ports can reduce engine efficiency.

By delivering twin sprays, directed at the intake ports without wetting the walls, Toyota gained closer control of the air-fuel ratio (which is critical when operating near the engine's lean limit), elimination of torque stumble and increased conversion efficiency in the three-way catalyst.

Mitsubishi has been using a Karman-vortex ultrasonic airflow meter for several years. Since 1985, Toyota has had a system in which the vortex is optically measured.

The vortex effect creates pressure, which is brought to bear against a thin metal mirror. The vibrations of the mirror are optically measured by a light-emitting diode and a phototransistor. This was adopted for the 1986 7M GTEU engine, a turbo-intercooled twin-cam inline three-liter six with four valves per cylinder, installed in the Toyota Supra coupe.

The L-Jetronic fuel injection system is divided into three groups, so that each pair of cylinders is supplied with fuel once per two revolutions. Fitted with twin-hole nozzles, it puts out 230 hp at 5600 rpm with a peak torque of 324 Nm at 4000 rpm.

Mazda Motors has successfully applied a fuel injection system based on the L-Jetronic to the Type 13 B Rotary (Wankel) engine of the RX-7 sports car. It went into production in February 1986.

The 13 B engine is a twin-rotor design with a cell volume of 654 cc. It has six intake ports and a mani-

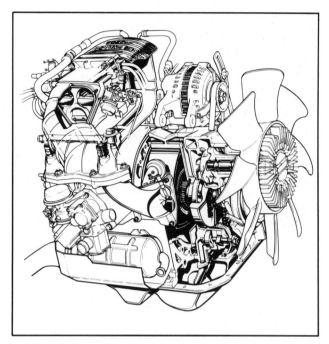

Mazda Motors adapted Bosch electronic fuel injection to the six-port Type 13 B rotary engine.

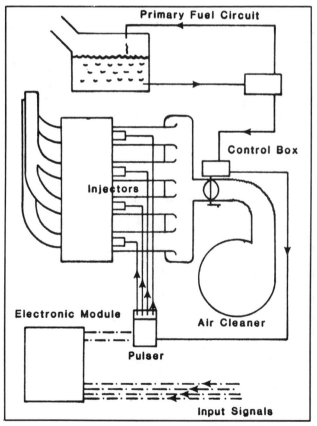

Schematic for the Pijet microprocessor-controlled fuel injection system, a low-cost solution for light economy cars.

157

folding system known as DEI (dynamic effect induction). The use of three lateral ports per chamber permits closer control over flow variations in the interest of providing optimum combustion efficiency and torque at all speeds, with smooth transitions. The torque curve is almost a straight line between 2500 and 5000 rpm.

Each chamber has two injectors. One, the primary injector, is mounted as close to the primary port as practicable. The other injector is installed a considerable distance upstream in the intake duct.

The rotor movement is phased so that there is a certain overlap between the intake phases. Whenever an intake phase ends, a pressure wave is set up, with a rebound effect that is timed to add its force to the airflow just as the next intake phase begins in the other chamber.

The primary injector delivers a carefully metered amount of fuel into the primary port. The secondary injectors are inactive during cruising, coasting, deceleration and idle operation. However, they provide mixture enrichment for acceleration in response to orders from the electronic control unit, which balances the change in throttle plate angle against other input such as rpm, vehicle speed, coolant temperature, mass airflow and air temperature.

The electronic control unit also receives oxygen content signals (Lambda-Sond) from the thermal reactor mounted in the forward section of the exhaust pipe. These are taken into account for continuous adjustment of the air-fuel ratio. Finally, the control unit times the spark and stabilizes the idle speed.

With two-stage turbocharging and intercooling, this engine puts out 182 hp at 6500 rpm, with a peak torque of 247 Nm at 3500 rpm.

Breaking the cost barrier

Dreams can sometimes turn into reality, but the dream of inventing a fuel injection system that costs no more than a carburetor seems a long way from reaching the hardware stage. Two systems created with that aim are the Pijet and a Volkswagen system.

The Pijet electronic fuel injection system was invented by Robert J. Gayler of Piper FM Limited. It was developed in 1980-83 as a low-cost device for light economy cars.

This system was based on the premise that engines prefer a constant air-fuel ratio over their entire load and speed range. It can be described as electronic multipoint carburetion with low-pressure port-mounted injector valves.

Fuel was pressurized to 30 psi by an eccentric roller-vane pump that was submerged in the tank and electrically driven on battery current. From the pump, the fuel flow went through a control box to a pulser. The control box served as a primary metering device, coordinating the throttle plate angle with the motion of a sleeve valve.

The second metering stage was handled by the pulser, which consisted of a solenoid-operated pintle

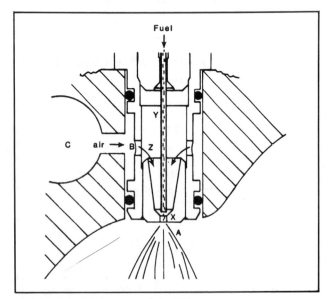

Air bleed fuel injectors of the Pijet system were mounted on the manifold, close by the cylinder head flange.

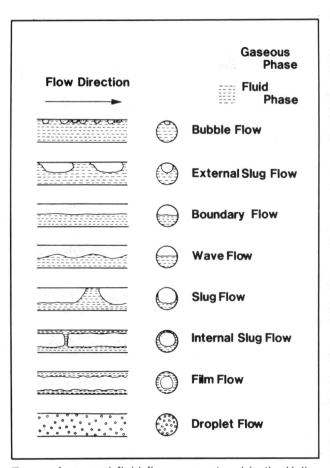

Types of gas and fluid flow encountered in the Volkswagen air-forced injection system.

valve that was spring-loaded to a closed position. The pintle was lifted from its seat when the core was magnetized, thus allowing a certain amount of fuel to pass through. The pulse width of the current was determined by the electronic control unit in accordance with input data from a number of sensors.

The pintle valve had very high oscillation frequency, with energized periods of between three and ten milliseconds. Frequency control depended exclusively on engine speed, and the injection period was controlled by the energizing pulse width.

The pulser was integrated with a fuel accumulator that regulated delivery pressure to a maximum of 10 psi. The accumulator was capable of maintaining that pressure on the overrun or during fuel shutoff.

The port-mounted injectors were air bleed valves that responded to an airflow along individual inlet tracts. Since this airflow was proportional to the breathing rate of one particular cylinder, fuel delivery was restricted to the opening period of the intake valve for that cylinder. This ensured a natural balance in the fuel distribution.

Experimental Pijet installations were tested on a number of cars with engines between 1100 cc and two liters displacement. Results corresponded to a change from carburetors to a typical electronic fuel injection system, in fuel economy terms as well as with respect to emission levels.

As a big buyer of K-Jetronic equipment from Bosch, Volkswagen had been busily exploring new ways to reduce the cost of the mechanical-hydraulic fuel injection system since 1975. Starting with the idea of using compressed air in place of mechanical pumping action, a group of Wolfsburg engineers developed the air-forced injection system in 1984-85. The experimental applications were made on the one-liter Volkswagen Polo engine.

How did it work? A small amount of the incoming airflow was drained off and ducted to the fuel supply line, where it enlarged the volume of the small fuel flow to a large mass of combustible mixture, ready for distribution to the individual cylinders. Ideally, this should be possible without the use of injectors, but owing to the huge variety of patterns that can occur whenever air and fuel are brought together, that proved to be a very difficult goal to reach.

Economy cars do not usually possess a built-in source of compressed air, and thus the system had to include its own compressor. After testing fifteen different types, the Volkswagen engineers decided in favor of a small piston-type compressor of about 8 cc derived from a refrigerator unit.

The first-generation design revealed serious problems regarding the size and selection of tubing for the connections between the pump and the mixture distributor, and from there to the inlet ports. The design of the mixture distributor itself brought a number of difficulties to the surface.

After many months, enough test experience had been obtained to permit the drawing up of a second-

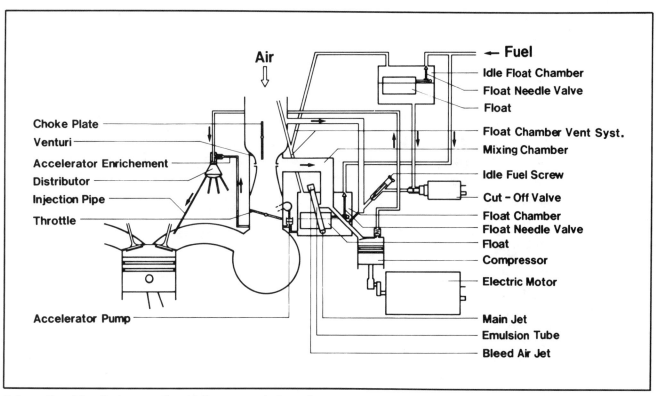

Schematic of the first-generation Volkswagen air-forced injection system.

generation system. Here, the fresh air passed over the choke plate for cold-starting and warm-up before reaching a venturi with a number of slots at the throat.

The venturi served the same purpose as the venturi in a typical carburetor. Its slots transmitted a pressure signal to the metering orifice between the metering rod and the metering jet (both derived from carburetor practice). The stroke of the rod was dependent on the throttle plate angle, and the jet registered the fuel level in the float chamber. Together, these accomplished the complete fuel-metering task.

The slots on the venturi leaked air to the mixing chamber through a passage with low pressure loss. They also formed air passages leading to the compressor, which passed about fifteen percent of the combustion airflow at idle speed. The compressor was driven by an electric motor. It had a capacity of about 400 cc per second at idle speed, and provided peak pressures up to three bar to assist in atomization of the fuel particles.

The mixing chamber had to be large to avoid an important drop in pressure. The bottom of the mixing chamber had the inlet port to the compressor on one side and an entry port for the fuel on the other side. Fuel supply was cut off by a valve during standstill and on overrun.

A separate accelerator pump had to be fitted, since the basic system was unable to respond to sudden loads from a low part-load situation. A normal diaphragm-type pump, actuated by a cam on the throttle shaft, was added, with its own injection channel leading direct to the mixture distributor.

Rig tests proved that the system automatically supplied a stoichiometric air-fuel ratio with only slight deviations. In terms of performance, exhaust emissions, cold- and hot-starts, fuel economy and drivability, it proved equivalent to established intermittent multipoint nonelectronic fuel injection systems (K-Jetronic).

Turbocharging and water injection

The current popularity of turbocharging has revived interest in water injection, either as part of the fuel injection system or as a separate system. Water possesses a high amount of latent heat, which makes it suitable for internal cooling of the combustion process. The water improves the engine's knock resistance and cools the exhaust gas, thereby permitting higher boost pressures while lowering the thermal stress on the turbine section.

By slowing down the burning rate, water reduces the pressure rise rate and the maximum cycle temperature. That cuts down on NO_x emissions. It cools the end gas, thereby preventing it from reaching critically high temperatures and causing self-ignition. Water and water-alcohol mixtures have proved able to prevent, to a great extent, the build-up of carbon deposits

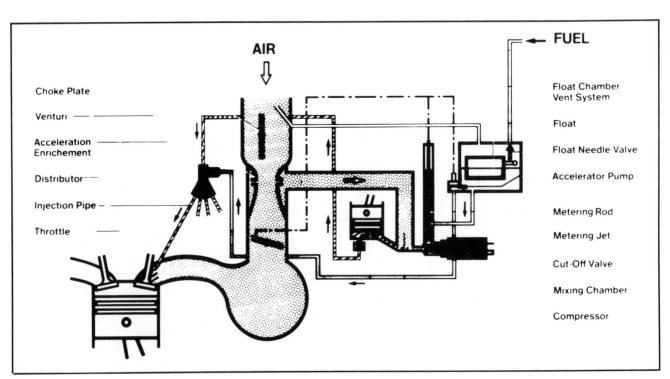

Schematic of the second-generation Volkswagen air-forced injection system.

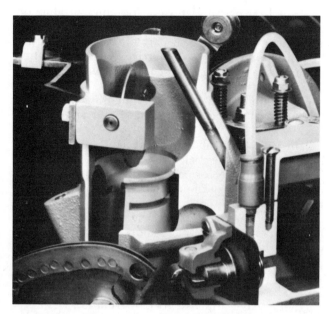
Detail view inside the Volkswagen air-forced injection system throttle body.

in an engine. They also protect valves and valve seats from burning, pitting and distortion. Water injection may permit changes in spark timing to get higher efficiency.

But water should not be injected into cold engines, where it cannot be fully vaporized and will increase corrosion in the exhaust system. It will also slow down the warm-up, worsening the emission level and wasting fuel.

Water injection was first tried by a Hungarian engineer named Benki before 1900. He was led to experiment with water injection from the realization that water can be used as an internal coolant to prevent preignition and detonation.

The first scientific tests with water injection were made in 1913 by Professor Bertram Hopkinson at Trinity College in Cambridge, England, on horizontal-cylinder industrial gas engines. About 1925, it became fairly common to find water injection as an antidetonation device on farm tractor semidiesel engines fitted with hot-bulb ignition.

Starting about 1927, when aircraft engine makers developed supercharged power units for speed and altitude records, water injection came into renewed focus. This line of research was given further impetus when Dr. Sanford Moss of General Electric perfected the turbocharger for aircraft engine installation.

In most water injection systems made before 1939, the water was injected into the air-fuel mixture before it entered the supercharger. Water and water-alcohol mixtures were injected into US Army Air Force, Royal Air Force and Luftwaffe aircraft engines during World War II. To assist in takeoff and increase flight speed on certain aircraft engines, cooling the combustion air by 22°C was found to give a four percent increase in power output.

After the war, civil aviation showed no interest in water injection, which retreated to the laboratories. In 1945, Arch Colwell of the GM Research laboratories

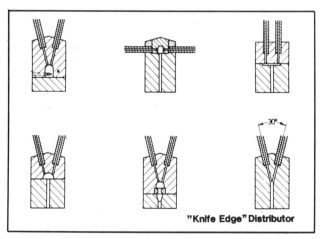
Six different configurations of fuel mixture distributors were tested with the Volkswagen air-forced injection system.

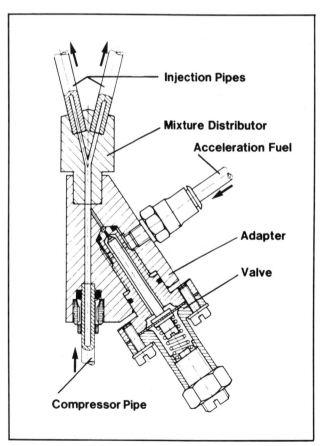
The Volkswagen air-forced injection system had an enrichment device for acceleration.

Mixture delivery pattern in the Volkswagen air-forced injection system as installed on the four-cylinder 1.3 liter Polo engine.

led a team that published a comprehensive report on water injection. In the 1950s, an Australian firm, Kleinig Products, began manufacturing the TT Mist-Master as an aftermarket water injection kit, but the auto industry saw no use for it.

In 1952, J. Jalbert, director of the Compagnie des Moteurs a Combustion in France, tested water injection and found that with sixty-five-octane gasoline, the compression ratio could be raised from 6:1 to 9 or 10:1.

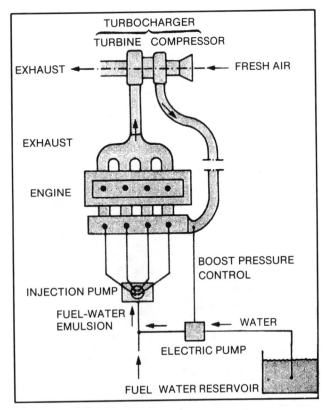

Water injection: One approach is to mix water into the fuel upstream of the injection pump.

The first use of water injection on a production model car had to wait until 1962, when Oldsmobile adopted it as an integral part of the turbocharger installation on the F-85 Jetfire (which breathed through a carburetor). Porsche tested water injection on a single-cylinder test engine in 1971-72, and about the same time, BMW was experimenting with water injection on turbocharged four- and six-cylinder engines.

In 1978-79, Saab engine specialists tested water injection as a potential means of improving turbocharged engine performance in the face of deteriorating fuel quality without increased risk of causing knock.

Water was injected into the manifold to reduce the temperature of the compressed air before it entered the cylinders. A sensor registered the boost pressure and, at a certain boost level, activated a separate water pump, which drew water from a small tank. The water was forced through a filter and into a magnetic valve attached to a distribution chamber. This valve regulated the flow to injector nozzles mounted on the intake manifold runners. The experimental 170 hp superturbo was based on the Saab 900 rally car, and it timed at speeds up to 210 kph.

In the spring of 1982, Ferrari adopted water injection (in addition to fuel injection) on the 126 C 2 Formula One racing engine, a turbocharged 1.5 liter V-6 delivering about 650 hp. The system was developed by Solar 77 of Rome in late 1981, and Ferrari secured a ten-year exclusive contract for it.

By 1983, however, practically all other turbocharged Formula One engines had been fitted with alternative water injection systems. That year, Peugeot and Renault used water injection in their rally car engines, and Volvo used it on its racing (Touring Car Class) sedans.

In the Ferrari 126 C installation, the fuel tanks were made smaller to accommodate a twenty-liter water tank, and the injection rate was controlled by turbocharger boost pressure. The water was blended with the gasoline in a mixing device, and peak exhaust gas temperatures sank from 900 to 850°C. With this method, the proportion of water was restricted to about six percent of the fuel flow. Pumping water from a separate tank into the inlet manifold permits water proportions up to fifteen percent.

The optimum water-fuel ratio is not a constant value. At high rpm, for example, the water proportion must be reduced. An engine that ran well with thirty parts water to 100 parts gasoline under full throttle at 1000 rpm would not tolerate more than four parts water to 100 parts gasoline at 4000 rpm.

More widespread use of water injection will not occur before some of the cost has been taken out of it. For instance, the water injection pump could be eliminated by draining off some manifold boost pressure (by means of a nonreturn valve) to pressurize the water tank. That would force water out the delivery pipe and into the entry tract of the turbocharger. But this method has the drawback of causing water evaporation in the compressor; superefficient intercooling would be required to again condense the water.

Two-stroke engines and pneumatic fuel injection

Two-stroke engines were once common in light economy cars and small utility vehicles. They disappeared primarily because they failed to meet emission control standards and also because of their excessive fuel consumption.

Now the two-stroke is back in the news, and Ford, Chrysler and General Motors are spending lots of money on two-stroke projects. In 1985, GM even began buying into an Australian firm, the Orbital Engine Company, whose founder, Frank Sarich, held some interesting two-stroke engine patents. (Until 1985, this firm was owned sixty percent by the Sarich Technology Trust and forty percent by Broken Hill Proprietary, a diversified mining giant).

The main attractions of the two-stroke engine are as follows:
• Ability to fire regularly over a wide range of load and speed
• Ability to maintain high mean effective pressure at high rpm

• Good balance
• Potential for suppression of fluid and friction losses
The auto industry is also interested in the two-stroke engine because of its mechanical simplicity and its implied promise of lower manufacturing cost, improved power-weight ratio and smaller power-package size.

The car makers have had plenty of time to look at the two-stroke. It is older than both the four-cycle gasoline engine and the diesel; J.J.E. Lenoir made a two-stroke gas engine in 1862, and N. A. Otto patented one in 1866. Today, this engine has extensive application only in outboard motors, light aircraft and motorcycles.

The assessment of Martin Ford-Dunn, a two-stroke engine specialist in the laboratories of Ricardo Consulting Engineers of Shoreham-on-Sea, Sussex, is likely to remain valid for an indefinite future. Among his qualifications are the complete design of an integral motorcycle-sidecar combination that has set a number of world speed records and is powered by a four-cylinder Ricardo two-stroke engine of Ford-Dunn's own concept.

"The main advantage of the two-stroke engine is that it is small and light," Ford-Dunn emphasizes. "It has few moving parts—in its simplest form, and it has low internal friction. Because the engine size will be smaller, the two-stroke engine promises lower hood lines. It can also save weight throughout the car."

But isn't it a notorious fuel waster? "It's *possible*," stresses Ford-Dunn, "to get *better* fuel mileage from a

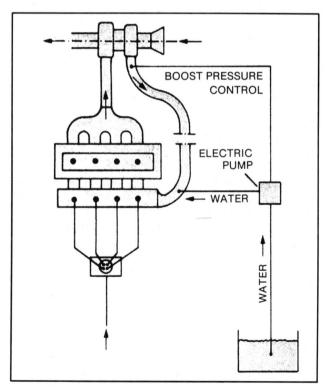

Water injection: Another way is to inject water into the manifold by means of a separate system, fully independent of the fuel injection, though metering must be coordinated.

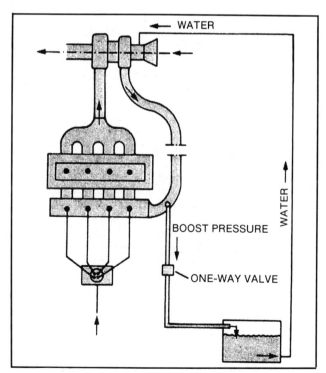

Water can be injected into the compressor mouth, thus relying on air pressure for its delivery into the cylinders.

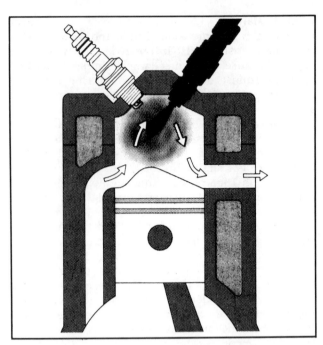

The principle of fuel injection for two-stroke engines is schematically explained in this sketch. Use of direct injection minimizes waste of fresh fuel down the exhaust pipe, with an analog drop in emission levels.

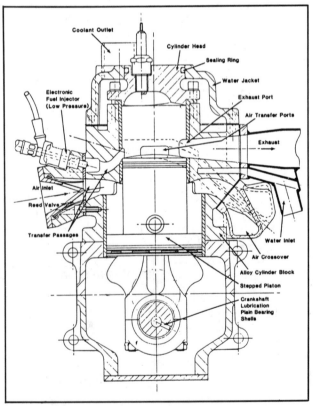

Latest version of the Templewall two-stroke engine has a low-pressure injector nozzle mounted in a transfer passage that feeds into the upper cylinder area.

two-stroke than you can from a four-stroke engine, but it's far from certain. It demands direct fuel injection, and a lot of work on the combustion chamber design."

Lined up against the two-stroke engine is a formidable obstacle: Its burning efficiency is lower than the four-stroke engine's, and improving it to the same level cannot be done without making mechanical design changes. Consequently, its cost-benefit ratio won't be as good as a four-stroke engine's.

The engine also has significant difficulties with lube oil. "The problem is the cylinder wall, how to keep the piston moving freely without risk of seizure," says Ford-Dunn. "You scavenge the engine to avoid mixing oil into the fresh charge. But there's a long way to go."

Would direct fuel injection alone be enough to enable the two-stroke engine to meet present and future emission control standards? "Low NO_x emissions are inherent in the design," explains Ford-Dunn. "HC [Hydrocarbon] can be tolerable—if the fuel cost is accepted. It will burn more fuel to meet the HC limits."

"Just look at the pressure diagram," Ford-Dunn continues. "It shows a series of nonfiring strokes. Then one that fires. The two-stroke engine in its present form has an irregular nature. With electronic controls, fuel supply could be cut off, just in time. Or it might be solved with stratified charge. If we decide to try, the control boys will have a field day."

When Ford-Dunn says "control boys," he means Ricardo's own specialists on electronic control systems, devices and software.

"If given an assignment to draw up a new two-stroke automobile engine today, its type and configuration would depend on the size and weight of the car, to a large extent. There are obvious three-cylinder benefits. We would begin by looking at multiples of three cylinders. For a 750 kg or smaller car, you can do better with fewer cylinders, say two working cylinders and one pumping cylinder. Scavenge blowing would assure adequate cylinder filling, if scavenge rates are made high enough. Or you could use a reciprocating blower—add 'slave' cylinders to provide the pumping. If that doesn't give an adequate scavenging rate, turbocharging could be the answer."

Many engineers, including Bernard Hooper, would argue that the two-stroke engine won't succeed as an automotive power unit until entirely novel concepts are developed. Hooper invented the Templewall stepped-piston two-stroke engine. (Templewall is a research organization created by former employees of the Norton Villiers Triumph motorcycle company.) The engine design dates from 1980. It exists as a 500 cc parallel twin for motorcycles and a one-liter ninety degree V-4 for cars, with a unique electronic fuel injection system.

With fuel injection in a two-stroke engine, there is no need to mix lube oil in the gasoline; thus, the engine's traditional bane with regard to emission control is eliminated. When the crankcase is divorced from breathing and compression duties, there is no bearing lubrication problem. A one-piece crankshaft, running in plain bearings, can be used.

The Templewall engine concept is based on a stepped piston that moves up and down in two concentric cylinder liners of different bore sizes (a small bore in the upper cylinder and a wide bore in the lower part). All ports are grouped around the step where the bore size changes. The wrist pin is mounted in the narrow-diameter part of the piston, just clear of the step to the wide diameter. The small end carries two compression rings, and the wide end only one.

How does this engine work? Fresh air is drawn into the wide-bore section on the piston's downstroke through a nonreturn reed valve in the inlet port. The two cylinders of the motorcycle twin engine are connected by crossover passages, so that the air is pumped from the wide-bore section of one cylinder to the small-bore section of the other cylinder. During the power stroke, the piston travel uncovers the transfer port outlet and fresh air enters for scavenging and the next compression.

Fuel is injected at low pressure in the initial period of the compression stroke. At the same time, the

exhaust port closes so as to retain the total fuel charge in the combustion chamber. The geometry of the transfer passages ensures a degree of charge stratification, which has the effect of keeping the rich portion of the charge far away from the exhaust port. This not only prevents heat loss but also assures the engine of an extreme lean-burn capability.

The low-pressure electronic fuel injection system was developed in cooperation with SU-Butec. It uses a fuel rail pressurized at 28 psi and a pressure regulator that ensures delivery at a mere 10 psi, which is enough to provide a rich mixture when needed.

The 500 cc motorcycle engine put out 37 hp at 5200 rpm. Specific fuel consumption under full load was recorded as 220 grams per hp-hr, which is fully competitive with four-stroke engines in that displacement class.

No proof has been given that the Orbital (Sarich) two-stroke engine can match the Templewall design for power density or fuel efficiency. In fact, Orbital's engine development effort seems to have been pushed into the background in favor of a pneumatic fuel injection system that is applicable to both two-stroke and four-stroke engines.

Pneumatics are coming into high fashion. In marine engineering, pneumatic circuits are used

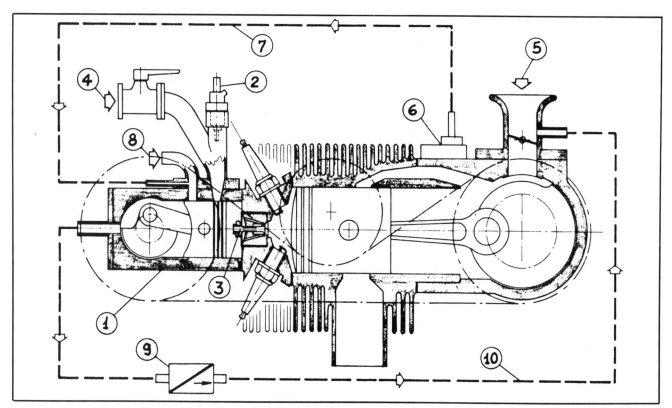

Piaggio experimental two-stroke engine with pneumatic fuel injection. 1-Piston-type injection system air compressor, 2-Electrically controlled fuel injector valve, 3-Automatic injector valve, 4-Injection air intake, 5-Scavenging air intake, 6-Lubrication oil pump, 7-Lube-oil circuit, 8-Emulsion air intake, 9-Nonreturn valve, 10-Emulsion circuit.

increasingly to replace hydraulics. In 1986, Renault adopted a pneumatic valve-closing system for its V-6 F-1 racing engines, extending the valve-crash limit by 1500 rpm or more.

The Orbital system works by an air blast that propels the fuel out of a distributor body to the cylinder ports. The air blast comes from a belt-driven compressor mounted on the engine. It is triggered by an electronic control system that provides timed and sequential injection under low pressure to each intake port. The blast air is mixed with the fuel, which is claimed to accelerate the atomization of the fuel.

The basic system was invented in 1983. Orbital Engine Company then converted a Suzuki two-stroke motorcycle engine to pneumatic fuel injection in 1986 and planned its own prototype engine for 1987.

Pneumatic fuel injection has been around a lot longer than some modern inventors realize. During the war years, 1942-45, J. Jalbert developed a pneumatic system for fuel injection that was suitable for both diesel and gasoline engines.

Jalbert was a graduate of the French Ecole Polytechnique and a marine engineer with the shipyards of the Loire. He pointed out that higher bmep (mean effective pressures) could be obtained with pneumatic injection than with mechanical injection. The construction of Jalbert's pneumatic injection consisted of a charging cylinder placed above each working cylinder. Each charging cylinder had its own piston and overhead crankshaft, and in a four-stroke engine it ran at half the speed of the power-producing crankshaft.

The injection cylinder was separated from the working cylinder by a poppet valve. An overrich air-fuel mixture was fed into the injection cylinder from a primitive carburetor. On its compression stroke, the injection cylinder piston built up enough pressure to force the injection valve open, admitting the overrich mixture into the working cylinder during the intake stroke, to mix with the fresh air.

Jalbert's system was tested on a 27.6 liter V-16 aircraft engine made by the French firm Ateliers et Chantiers de la Loire. This engine put out 600 hp at 2650 rpm, with a minimum specific fuel consumption of 178 grams per hp-hr. In 1948, the same system was adopted for a big V-12 diesel engine used in French railway locomotives.

The latest advances in pneumatic fuel injection stem from a research program undertaken in 1982-83 by Piaggio S.A. of Pontedera near Pisa, Italy, maker of the Vespa (now Cosa) motor scooter and a line of three-wheeled commercial vehicles (ApeCar). Some

financial support for the project came from the National Research Council in Rome.

The work at Piaggio was carried out by a team led by Ing. Giovanni Batoni, assisted by Roberto Gentili and I. di Gangi of the Energy Department at the University of Pisa. It was a stratified charge concept, drawing on the experience of S. Yui and S. Onishi (SAE paper number 690468), and G. Yamagishi, T. Sato and H. Iwasa (SAE paper number 720195); and on the proposals of Jacques Pichard (Moteur Moderne). Pichard's theories were outlined in the protocol from the 4th International Symposium on Automotive Propulsion Systems, Volume III (1977).

The engine was based on the production model single-cylinder unit that powered the Vespa 200 Rally motor scooter, but with a new cylinder axis and provision for twin spark plugs. The breathing system, with crankcase compression of the fresh air, was modified in detail only.

The cylinder had a bore of 66.5 mm and the crankshaft gave a stroke of 57 mm, which makes for a displacement of 197.9 cc. The compression ratio was high (10.5:1), and the two spark plugs were fired by the same electronic pickup. The fuel injection pump was a piston-type air compressor with a mechanical V-belt driven from the crankshaft. Fuel was metered by an electronic control unit to an injector mounted on the compressor body, so as to begin the mixture preparation outside the cylinder. The automatic injection nozzle was opened by air pressure overcoming a spring-load.

The injection nozzle spring was preset to open at four bar pressure. The compressor cylinder dimensions were 43 mm bore and 21 mm stroke (30.5 cc). Maximum output was 11 hp at 5000 rpm (which is equivalent to 56 hp per liter!) and the maximum torque reached 15.9 Nm at 4000 rpm. Specific fuel consumption was 248 grams per hp-hr (0.546 pound) at full power and 235.3 grams per hp-hr (0.519 pound) at peak torque under full load.

Emission tests were made for unburned hydrocarbon, carbon monoxide and carbon dioxide. Hydrocarbon levels varied between 3.2 and 10.1 grams per kilowatt-hour and carbon monoxide levels ranged from ten to thirty-one grams per kilowatt-hour.

The men responsible for this project concluded that a two-stroke engine using these principles could be developed to match the pollution levels of a four-stroke unit, while giving superior specific output, especially under part-load conditions. In terms of fuel economy, however, the system still has some way to go.

Appendix

The following is a survey of fuel injection systems fitted on 1988-model cars.

Bosch K-Jetronic

Make	Model	Number of cylinders	Displacement in cc	HP(DIN)@RPM	Torque@RPM	Remarks
Audi	Quattro	5	2226	200@5800	270 Nm@3000	Turbo-intercooled
Eagle	Premier	V-6	2975	152@5000	232 Nm@3750	Peugeot-Renault-Volvo engine
Ferrari	GTB, GTS Turbo	V-8	1991	254@6500	329 Nm@4100	Turbo-intercooled
	Mondial	V-8	3186	270@7000	304 Nm@5500	
Lamborghini	Jalpa	V-8	3485	255@7000	314 Nm@3500	
	Countach	V-12	5167	455@7000	500 Nm@5200	
Lancia	Thema	V-6	2849	150@5750	240 Nm@2700	Peugeot-Renault-Volvo engine
Porsche	911 Turbo	Flat-6	3299	300@5500	430 Nm@4000	Turbo-intercooled
Renault	25	V-6	2849	160@5400	235 Nm@2500	Peugeot-Renault-Volvo engine
Saab	900 i	4	1985	118@5500	167 Nm@3000	
	900 Turbo	4	1985	155@5000	235 Nm@3000	Turbocharged
Volvo	240	4	2316	133@5400	195 Nm@3600	
	240 Turbo	4	2127	155@5500	240 Nm@3750	Turbocharged
	750 GLE	V-6	2849	170@5400	240 Nm@4800	Peugeot-Renault-Volvo engine

Bosch KE-Jetronic

Make	Model	Number of cylinders	Displacement in cc	HP(DIN)@RPM	Torque@RPM	Remarks
Audi	80 1.8 S	4	1781	90@5400	140 Nm@3350	Catalytic converter, 9.0:1 compression
	80 1.9 E	4	1847	113@5800	160 Nm@3400	Catalytic converter, 10.5:1 compression
	90 and 100	5	1994	115@5400	172 Nm@4000	Catalytic converter, 10.1:1 compression
	100	5	2309	136@5700	190 Nm@4500	Also available in the 90 2.3 E and 90 quattro
	100 2.2 Turbo	5	2226	165@5500	240 Nm@3000	Turbo-intercooled, 7.8:1 compression, catalytic converter
Bentley	Mulsanne S	V-8	6750	n/a	n/a	Also available in the Continental and Rolls-Royce Silver Spur and Silver Spirit
	Turbo R	V-8	6750	n/a	n/a	Turbocharged
Ferrari	328 GTB/GTS	V-8	3186	270@7000	304 Nm@5500	
	412	V-12	4943	340@6000	451 Nm@4200	
	Testarossa	Flat 12	4942	390@6300	490 Nm@4500	
Ford	Escort	4	1598	132@5750	181 Nm@2750	Turbocharged
	Orion	4	1598	90@5800	123 Nm@4600	
Lancia	Thema 8.32	V-8	2927	215@6750	285 Nm@4500	Ferrari engine, KE3-Jetronic
Lotus	Esprit Turbo	4	2174	218@6000	270 Nm@5000	Turbocharged
Mercedes-Benz	190 E	4	1997	118@5100	172 Nm@3500	
	230 E	4	2299	132@5100	198 Nm@3500	Also available in the 190 E 2.3
	260 E and SE	6	2599	160@5800	220 Nm@4600	Also available in the 190 E 2.6

Bosch KE-Jetronic

Make	Model	Number of cylinders	Displacement in cc	HP(DIN)@RPM	Torque@RPM	Remarks
	300 E and SE	6	2962	180@5700	255 Nm@4400	
	420 SE	V-8	4196	224@5400	325 Nm@4000	
	500 SE	V-8	4973	252@5200	390 Nm@3750	
	560 SEL	V-8	5547	279@5200	430 Nm@3750	
Volkswagen	Golf GTI 16 V	4	1781	139@6100	168 Nm@4250	4 valves per cyl., also available in the VW Jetta and Scirocco

Bosch Mono-Jetronic

Make	Model	Number of cylinders	Displacement in cc	HP(DIN)@RPM	Torque@RPM	Remarks
Audi	100 1.8	4	1781	90@5500	142 Nm@3250	Catalytic converter 9.0:1 compression
Chevrolet	Monte Carlo	V-6	4300	147@4200	305 Nm@2000	
Fiat	Uno	4	999	45@5250	75 Nm@3250	Catalytic converter 9.0 to 9.5:1 compression, also available in the Fiat Panda and Autobianchi Y-10
	Tipo	4	1580	90@6250	123 Nm@4250	Catalytic converter 9.3:1 compression
Toyota	Corona	4	1838	150@5600	149 Nm@2800	
	Camry & Vista	4	1832	85@5200	142 Nm@3000	
Volkswagen	Passat	4	1781	90@5500	142 Nm@3000	

VW-Bosch Digijet

Make	Model	Number of cylinders	Displacement in cc	HP(DIN)@RPM	Torque@RPM	Remarks
Volkswagen	Polo	4	1272	55@5200	97 Nm@3000	Catalytic converter

VW-Bosch Digifant

Make	Model	Number of cylinders	Displacement in cc	HP(DIN)@RPM	Torque@RPM	Remarks
Volkswagen	Polo	4	1272	115@6000	148 Nm@3600	G 40 spiral compressor
	Golf GTI	4	1781	112@5500	157 Nm@3100	Also available in the Jetta
	Passat	4	1781	107@5400	154 Nm@3800	

Bosch L-Jetronic

Make	Model	Number of cylinders	Displacement in cc	HP(DIN)@RPM	Torque@RPM	Remarks
Alfa Romeo	75 6V 3.0	V-6	2959	188@5800	250 Nm@4500	
Autobianchi	Y 10 1.3 i.e.	4	1297	74@5750	100 Nm@3250	Catalytic converter 9.2:1 compression
Bertone	X-1/9	4	1499	76@5500	108 Nm@3000	Fiat engine, catalytic converter, 8.5:1 compression
Bitter	SC	6	2969	180@5800	243 Nm@4500	Opel-based engine
		6	3849	210@5100	321 Nm@3400	Opel-based engine
Buick	Le Sabre	V-6	3791	165@5200	285 Nm@2000	Also available on Buick Riviera, Electra and Reatta
Chevrolet	Corvette	V-8	5733	245@4000	454 Nm@3200	
Daihatsu	Charada	3	993	54@5200	80 Nm@3600	Catalytic converter
Fiat	Uno	4	1302	65@5600	100 Nm@3000	Catalytic converter 8.5:1 compression
	Regata	4	1499	75@5500	108 Nm@3000	Catalytic converter 8.5:1 compression
Ford	Escort	4	1859	112@5400	156 Nm@4200	US version
	Sierra	4	1993	115@5000	160 Nm@4000	Cologne-built version
Honda	Civic	4	1590	109@6300	135 Nm@5200	PGM-FI, 4 valves per cyl.
	Civic and CRX	4	1590	130@6800	143 Nm@5700	PGM-FI, 4 valves per cyl.
	Accord	4	1958	134@6000	171 Nm@5000	PGM-FI, 4 valves per cyl.
	Prelude	4	1958	145@6000	175 Nm@4500	PGM-FI, 4 valves per cyl.
	Legend	V-6	2494	173@6000	217 Nm@5000	PGM-FI, 4 valves per cyl.
	Legend	V-6	2675	180@6000	226 Nm@4500	PGM-FI, 4 valves per cyl.
Isuzu	Aska	4	1995	150@5400	226 Nm@3000	Made by Hitachi under Bosch license, turbo-intercooled, also available in the Piazza

Bosch L-Jetronic Make	Model	Number of cylinders	Displacement in cc	HP(DIN)@RPM	Torque@RPM	Remarks
	Piazza	4	1950	120@5800	162 Nm@4000	Made by Hitachi under Bosch license
Jaguar	XJ-6	6	3590	224@5000	337 Nm@4000	Also available in the Daimler 3.6 and Jaguar XJ-S. Supplied by Lucas
	XJ V-12	V-12	5343	295@5500	432 Nm@3250	Also available in the Daimler Double-Six, supplied by Lucas
Mazda	121	4	1290	88@7000	98 Nm@4500	4 valves per cyl.
	323	4	1323	68@5500	102 Nm@3500	Catalytic converter
	323	4	1598	86@5000	123 Nm@2500	
	323 Turbo	4	1598	140@6000	185 Nm@3000	Turbocharged
	Etude	4	1598	110@6500	132 Nm@4500	4 valves per cyl.
	626	4	1789	97@5500	143 Nm@4500	
	626	4	2184	115@5000	180 Nm@3000	
	626	4	2184	147@4300	258 Nm@3500	Turbo-intercooled, 3 valves per cyl.
	626	4	1998	140@6000	172 Nm@5000	4 valves per cyl.
	929	4	2184	115@5000	180 Nm@3000	
	929	V-6	1997	110@5500	168 Nm@4000	
	929	V-6	1997	145@5000	231 Nm@2500	Turbocharged, 3 valves per cyl.
	929	V-6	2954	160@5500	235 Nm@4000	3 valves per cyl.
	RX 7	RC2	2616	150@6500	182 Nm@3000	Wankel engine
	RX 7	RC 2	2616	185@6500	245 Nm@3500	Turbocharged Wankel
MG	Maestro	4	1994	117@5500	182 Nm@2800	Supplied by Lucas
Mitsubishi	Mirage/Colt	4	1468	82@5500	124 Nm@3000	Made by Mitsubishi under Bosch license
	Mirage/Colt	4	1595	125@6500	137 Nm@5200	Made by Mitsubishi under Bosch license
	Mirage/Colt	4	1595	145@6000	206 Nm@2500	Made by Mitsubishi under Bosch license, turbo-intercooled
	Sigma/Eterna	V-6	1998	105@5000	158 Nm@4000	Made by Mitsubishi under Bosch license
	Sigma/Eterna	V-6	1997	125@5500	202 Nm@3000	Made by Mitsubishi under Bosch license, turbocharged
	Sigma/Eterna	V-6	1997	200@6000	280 Nm@3500	Made by Mitsubishi under Bosch license, turbocharged with intercooling
	Sapporo	4	2351	124@5000	189 Nm@3500	Made by Mitsubishi under Bosch license
	Galant	4	1755	94@5000	143 Nm@3500	Made by Mitsubishi under Bosch license
	Galant	4	1997	112@5500	160 Nm@4500	Made by Mitsubishi under Bosch license
	Galant	4	1997	140@6000	172 Nm@5000	Made by Mitsubishi under Bosch license, 4 valves per cyl.
	Galant	4	1997	205@6000	294 Nm@3000	Made by Mitsubishi under Bosch license, 4 valves per cyl. turbo-intercooled
	Starion	4	2555	155@5000	284 Nm@2500	Made by Mitsubishi under Bosch license, turbocharged
	Starion	4	2555	191@5000	318 Nm@2500	Made by Mitsubishi under Bosch license, turbocharged with intercooler
	Debonair V	V-6	1998	105@5000	158 Nm@4000	Made by Mitsubishi under Bosch license
	Debonair V	V-6	1998	150@5500	221 Nm@3000	Made by Mitsubishi under Bosch license plus Roots blower with intercooling
Nissan	Sunny	4	1488	70@5600	121 Nm@2800	
	Sunny	4	1598	110@6400	130 Nm@4800	4 valves per cyl.
	Silvia	4	1809	135@6000	196 Nm@3600	Turbocharged
	Bluebird	4	1809	145@6400	201 Nm@4000	Turbocharged, 4 valves per cylinder
	Maxima	V-6	2960	154@5200	227 Nm@3600	Also available in the Leopard
	Silvia	V-6	2960	167@5200	228 Nm@3200	
	Bluebird	4	1809	175@6400	226 Nm@4000	Turbo-intercooled, 4 valves per cyl.

Bosch L-Jetronic (continued)

Make	Model	Number of cylinders	Displacement in cc	HP(DIN)@RPM	Torque@RPM	Remarks
	Skyline	6	1998	115@5600	167 Nm@4000	Also available in the Leopard and Laurel
	Skyline 2.0 24 V	6	1998	150@6400	181 Nm@5200	4 valves per cyl., also available in the Laurel
	Skyline GTS	6	1998	190@6400	240 Nm@4800	Turbo-intercooled, 4 valves per cyl.
	Leopard	V-6	2960	185@6000	245 Nm@4400	4 valves per cyl.
	Laurel	6	1998	175@6400	226 Nm@3600	Turbo-intercooled, 4 valves per cyl.
	Cedric/Gloria	V-6	1998	125@6000	167 Nm@3200	
	Cedric/Gloria	V-6	1998	185@6800	216 Nm@4800	Turbocharged
	Cedric/Gloria	V-6	2960	166@5200	236 Nm@4400	Also available in the Z-car
	Cedric/Gloria	V-6	2960	195@5200	310 Nm@3600	Turbocharged, also available in the Z-car
	Cedric/Gloria	V-6	2960	200@6000	260 Nm@4400	4 valves per cyl.
	Cedric/Gloria	V-6	2960	255@6000	343 Nm@3200	Turbocharged, 4 valves per cyl.
	Fairlady Z	6	1998	210@6400	265 Nm@3600	Turbocharged, 4 valves per cyl.
	Fairlady Z	V-6	2960	190@6000	246 Nm@4000	ECCS with NICS, 4 valves per cyl.
	President	V-8	4414	200@4800	338 Nm@3200	ECCS induction system
Opel	Kadett	4	1796	112@5600	158 Nm@3000	L-3.1 Jetronic, also available in the Ascona
	Omega	4	1796	115@5600	160 Nm@4600	L-3.1 Jetronic, also available in the Ascona
Peugeot	104 GTI	4	1580	115@6250	134 Nm@4000	
	505	4	2165	122@5750	180 Nm@4250	
	505	4	2165	167@5200	260 Nm@3000	Turbocharged
Porsche	944	4	2479	160@5900	210 Nm@4500	Also available in the 924 S
	944 S	4	2479	190@6000	230 Nm@4300	4 valves per cyl.
	944 Turbo	4	2479	220@5800	330 Nm@3500	Turbo-intercooled
	944 Turbo S	4	2479	250@6000	350 Nm@4000	Turbo-intercooled
	911 Carrera	Flat 6	3164	231@5900	284 Nm@4800	
Rover	216	4	1598	104@6000	137 Nm@3500	Supplied by Lucas
	Montego	4	1994	113@5600	182 Nm@2800	Supplied by Lucas
	800	4	1996	140@6000	179 Nm@4500	Supplied by Lucas
	800	V-6	2494	173@6000	217 Nm@5000	Honda PGM-FI
	800	V-6	2675	177@6000	228 Nm@4500	Honda PGM-FI
Seat	Ibiza 1.5	4	1461	100@5900	128 Nm@4700	LU-Jetronic, engine designed by Porsche, also available in the Seat Malaga
Subaru	Leone 1.8	Flat 4	1782	105@6000	142 Nm@3600	
	Leone 1.8 Turbo	Flat 4	1782	120@5200	183 Nm@3200	Turbocharged, also available in the XT coupe
	XT coupe	Flat 6	2672	150@5200	211 Nm@4000	
Suzuki	Cultus/Swift	3	993	82@5500	118 Nm@3500	Turbocharged
	Swift GTi	4	1299	101@6600	113 Nm@5500	4 valves per cyl.
Toyota	Corolla	4	1498	94@6000	129 Nm@4400	4 valves per cyl.
	Corolla	4	1587	120@6600	142 Nm@5200	4 valves per cyl.
	Corolla	4	1587	145@6400	186 Nm@4400	Roots blower with intercooling, also available in the MR 2
	Tercel	4	1456	110@5600	169 Nm@3200	
	Celica	4	1587	124@6600	142 Nm@5000	4 valves per cyl., also available in the MR 2 and Corona
	Celica	4	1998	120@5600	169 Nm@4400	Also available in the Camry and Vista
	Celica	4	1998	140@6200	172 Nm@4800	Also available in the Camry and Vista
	Celica Turbo	4	1998	185@6000	240 Nm@4000	Turbocharged
	Camry	V-6	1993	140@6000	174 Nm@4600	4 valves per cyl.
	Cressida	6	1988	130@5400	172 Nm@4400	
	Cressida	6	1988	140@6200	173 Nm@4000	Also available in the Crown, Soarer and Supra
	Cressida	6	1988	185@6200	235 Nm@3200	Twin turbochargers, also available in the Crown

Bosch L-Jetronic (continued)

Make	Model	Number of cylinders	Displacement in cc	HP(DIN)@RPM	Torque@RPM	Remarks
	Cressida	6	2759	170@5600	230 Nm@4600	
	Crown	6	1988	105@5200	157 Nm@4000	Also available in the Soarer and Supra
	Crown	6	1988	160@6000	206 Nm@4000	Roots blower
	Supra Turbo	6	2954	230@5600	324 Nm@4000	Turbo-intercooled
	Century	V-8	3995	165@4400	289 Nm@3600	
Volkswagen	Passat	4	1595	75@5200	120 Nm@2700	
	Passat	4	1984	136@5800	180 Nm@4400	

Bosch LE-Jetronic

Make	Model	Number of cylinders	Displacement in cc	HP(DIN)@RPM	Torque@RPM	Remarks
Alfa Romeo	Sprint-Quadri-foglio verde	4	1717	118@5800	147 Nm@3500	LE 3.2 Jetronic
	75 1.8 Turbo	4	1779	155@5800	226 Nm@2600	LE 2 Jetronic, turbocharged
	164 Turbo	4	1995	175@5250	265 Nm@2500	LE 2 Jetronic, turbocharged
Bitter		6	2969	177@5600	240 Nm@4400	LE 2 Jetronic, Opel engine
BMW	316 i	4	1787	102@5800	140 Nm@4500	Catalytic converter, 8.2:1 compression
Citroen	Visa GTi	4	1580	115@6250	133 Nm@4000	LE 2 Jetronic
	BX 19 GTi	4	1905	125@5500	175 Nm@4500	LE 3 Jetronic
	CX 25	4	2500	138@5000	211 Nm@4000	LE 2 Jetronic
	CX 25 Turbo	4	2500	168@5000	294 Nm@3250	LE/LU 2 Jetronic
Fiat	Uno Turbo	4	1301	105@5750	147 Nm@3200	LE 2 Jetronic
	Croma Turbo	4	1995	155@5250	235 Nm@2350	LE 2 Jetronic
Ford	Sierra XR 4x4	V-6	2792	150@5700	216 Nm@3800	
	Scorpio	V-6	2394	130@5800	193 Nm@3000	
	Scorpio	V-6	2935	150@5700	233 Nm@3000	
Lancia	Thema	4	1995	120@5250	167 Nm@3300	
	Thema 2000 i.e.	Turbo 4	1995	165@5500	255 Nm@2500	LE 2 Jetronic Turbocharged
Morgan	Plus-8	V-8	3532	193@5280	298 Nm@4000	Supplied by Lucas Rover engine
Opel	Corsa GSi	4	1598	100@5500	135 Nm@3400	
	Manta GT	4	1979	110@5400	162 Nm@3400	
	Senator	6	2490	140@5200	205 Nm@4200	LE 2 Jetronic
Peugeot	205 GTI	4	1905	130@6000	165 Nm@4750	LE 2 Jetronic, also available in 309 GTI
	405	4	1905	107@6000	140 Nm@3000	Catalytic converter
	405 SRI	4	1905	125@5500	175 Nm@4500	LE 3 Jetronic
Volvo	360 GLT	4	1986	115@5700	160 Nm@4200	

Bosch LH-Jetronic

Make	Model	Number of cylinders	Displacement in cc	HP(DIN)@RPM	Torque@RPM	Remarks
Jaguar	XJ-6	6	2919	167@5600	239 Nm@4000	
Peugeot	505	V-6	2849	170@5600	235 Nm@4250	Peugeot-Renault-Volvo engine
Porsche	928 S4	V-8	4957	320@6000	431 Nm@3000	4 valves per cyl.
Saab	900 i	4	1985	125@5500	170 Nm@3000	Catalytic converter, 10:1 compression
	900 Turbo 16 and 9000 Turbo 16	4	1985	175@5300	273 Nm@3000	Turbo-intercooled
	9000 i	4	1985	130@5500	173 Nm@3000	
Volvo	480 ES turbo	4	1721	120@5400	175 Nm@4800	Turbo-intercooled, Renault engine

Bosch Motronic

Make	Model	Number of cylinders	Displacement in cc	HP(DIN)@RPM	Torque@RPM	Remarks
Alfa Romeo	75 Twin Spark	4	1962	148@5800	191 Nm@4000	Motronic ML 4.1, 8 spark plugs
	Alfa 164	4	1962	148@5800	186 Nm@4000	Motronic ML 4.1, 4 spark plugs
	Alfa 164	V-6	2959	192@5600	245 Nm@3000	
Bitter		6	2969	156@5400	230 Nm@4000	Based on Opel engine, catalytic converter
		6	3849	200@5100	298 Nm@3400	Based on Opel engine, catalytic converter

Bosch Motronic

Make	Model	Number of cylinders	Displacement in cc	HP(DIN)@RPM	Torque@RPM	Remarks
BMW	318 i	4	1796	113@5500	162 Nm@4250	Catalytic converter, 8.8:1 compression
	320 i	4	1991	129@6000	164 Nm@4300	Catalytic converter, 8.8:1 compression, also available 520 i
	325	6	2693	129@4800	231 Nm@3250	US version, catalytic converter
	325 i and 525 i	6	2494	170@5800	222 Nm@4300	Catalytic converter, 8.8:1 compression
	M3	4	2302	195@6750	230 Nm@4750	Catalytic converter, 10.5:1 compression
	M3	4	2302	200@6750	240 Nm@4750	Catalytic converter, 10.5:1 compression
	Z 1	6	2494	170@5800	222 Nm@4300	Catalytic converter, 8.8:1 compression
	530 i and 730 i	6	2986	188@5800	260 Nm@4000	Catalytic converter, 9.0:1 compression
	535 i, 735 i and 635 CSi	6	3430	211@5700	305 Nm@4000	Catalytic converter
	M 635 S	6	3453	260@6500	330 Nm@4500	Catalytic converter, 9.8:1 compression
	750 i	V-12	4988	300@5200	450 Nm@4100	Catalytic converter, 8.8:1 compression
Citroen	BX 19 GTI	4	1905	160@6500	181 Nm@5000	4 valves per cyl.
Opel	Kadett GT 2.0 i	4	1998	129@5600	180 Nm@5600	Also available in the Ascona GT Sport
	Kadett GSi 16V	4	1998	150@6000	196 Nm@4800	4 valves per cyl.
	Omega 3.0 i	6	2969	177@5600	240 Nm@4400	Also available in the Senator
Peugeot	405 MI 16	4	1905	160@6500	181 Nm@5000	4 valves per cyl., ML4.1
Porsche	959	Flat 6	2850	450@6500	500 Nm@5500	4 valves per cyl., twin turbochargers intercooling

Bendix single-point injection

Make	Model	Number of cylinders	Displacement in cc	HP(DIN)@RPM	Torque@RPM	Remarks
Eagle	Premier	4	2484	113@4750	193 Nm@2500	
Jeep	Wrangler	4	2484	119@5000	183 Nm@3500	
	Cherokee	4	2484	123@5000	191 Nm@3500	
Renault	5	4	1397	60@4750	100 Nm@3000	Catalytic converter, 9.0:1 compression

Bendix multi-point injection

Make	Model	Number of cylinders	Displacement in cc	HP(DIN)@RPM	Torque@RPM	Remarks
MVS	Venturi	V-6	2458	200@5750	290 Nm@2500	Peugeot-Renault-Volvo engine Turbo-intercooled
Renault	5 GTE	4	1721	75@5500	129 Nm@3500	Also available in the 9 GLE and 11 GLE, catalytic converter, 9.5:1 compression
	21	4	1995	120@5500	168 Nm@4500	
	21 Turbo	4	1995	175@5200	270 Nm@3000	Turbocharged
	25	4	2165	126@5250	189 Nm@2750	
	25 Turbo	V-6	2458	182@5500	281 Nm@3000	Turbo-intercooled Peugeot-Renault-Volvo engine
	Alpine Turbo	V-6	2458	200@5750	290 Nm@2500	Turbo-intercooled Peugeot-Renault-Volvo engine
Volvo	480 ES	4	1721	109@5800	140 Nm@4000	Renault engine

Ford single-point injection

Make	Model	Number of cylinders	Displacement in cc	HP(DIN)@RPM	Torque@RPM	Remarks
Ford	Escort (USA)	4	1859	91@4600	144 Nm@3400	
	Taurus	4	2498	91@4400	176 Nm@2600	

Ford Multi-point injection

Make	Model	Number of cylinders	Displacement in cc	HP(DIN)@RPM	Torque@RPM	Remarks
Ford	Tempo	4	2300	99@4400	168 Nm@2200	Also available in the Mercury Topaz
	Tempo	4	2300	101@4400	176 Nm@2600	Also available in the Topaz XR 5 and LS Sport
	Mustang	4	2300	91@3800	176 Nm@2800	
	Mustang	V-8	4942	228@4200	407 Nm@3200	
	Taurus	V-6	2979	142@4800	217 Nm@3000	Also available in the Mercury Sable
	Thunderbird	V-6	3791	142@3800	292 Nm@2200	Also available in the Mercury Cougar
	Thunderbird	4	2300	152@4400	271 Nm@3000	Turbocharged
	Thunderbird	4	2300	193@4400	326 Nm@3000	Turbocharged
	LTD	V-8	4942	152@3200	366 Nm@2000	Also available in the Mercury Cougar
Lincoln	Continental	V-6	3791	142@3800	292 Nm@2200	
	Town Car	V-8	4942	152@3200	366 Nm@2000	Ex-Versailles
	Mark VII	V-8	4942	228@4200	407 Nm@3200	

Chrysler single-point injection

Make	Model	Number of cylinders	Displacement in cc	HP(DIN)@RPM	Torque@RPM	Remarks
Chrysler	Le Baron GTS	4	2213	177@5200	276 Nm@3200	Turbo-intercooled, also available in the Dodge Lancer and Plymouth Sundance
	Le Baron	4	2213	98@5200	166 Nm@3200	Also available in the Dodge Omni, Daytona, Shadow, Lancer, Aries and 600, Plymouth Caravelle and Sundance
	Le Baron	4	2501	100@4800	184 Nm@2800	Also available in the Dodge Daytona, Lancer, 600, and Dynasty
	TC by Maserati	4	2213	208@5500	298 Nm@3600	Turbocharged, 4 valves per cyl.
	New Yorker	V-6	2972	138@4800	228 Nm@2800	Mitsubishi engine, also available in the Dodge Dynasty

Chrysler multi-point injection

Make	Model	Number of cylinders	Displacement in cc	HP(DIN)@RPM	Torque@RPM	Remarks
Chrysler	ES Turbo	4	2213	148@5200	280 Nm@2400	Turbocharged, also available in the Chrysler LeBaron, Dodge Shadow, Daytona, Lancer, Plymouth Caravelle and Sundance

Rochester throttle body injection

Make	Model	Number of cylinders	Displacement in cc	HP(DIN)@RPM	Torque@RPM	Remarks
Buick	Skyhawk	4	1998	97@4800	160 Nm@3600	Also available in the Oldsmobile Firenza
	Skyhawk	4	1998	102@5200	176 Nm@2800	
	Skyhawk	4	2471	99@4800	183 Nm@3200	Also available in the Chevrolet Celebrity, Pontiac 6000 and Grand Am, Buick Century and Oldsmobile Cutlass Ciera
Cadillac	Seville	V-8	4467	155@4200	328 Nm@2800	Also available in the 60 Special, de ville, Fleetwood and Eldorado
	Allante	V-8	4087	172@4300	319 Nm@3200	
Chevrolet	Cavalier	4	1991	90@5600	147 Nm@3200	Also available in the Corsica and Beretta
	Caprice	V-6	4300	140@4200	305 Nm@2000	
Opel	Kadett S	4	1297	60@5600	96 Nm@3400	
Pontiac	Le Mans	4	1598	75@5600	122 Nm@2800	Opel-designed engine, made by Daewoo in Korea

Rochester multipoint injection

Make	Model	Number of cylinders	Displacement in cc	HP(DIN)@RPM	Torque@RPM	Remarks
Buick	Skylark	4	2260	152@5200	217 Nm@4000	Olds Quad-4 engine, also available in the Oldsmobile Cutlass Calais, Pontiac Grand Am
	Skylark	V-6	2966	127@4900	203 Nm@2400	Also available in the Buick Century and Oldsmobile Cutlass Calais
	Century	V-6	3791	152@4400	271 Nm@2000	Also available in the Buick LeSabre and Oldsmobile Cutlass Ciera
	Regal	V-6	2838	127@4500	217 Nm@3600	Also available in the Pontiac 6000, Oldsmobile Cutlass Ciera, Cadillac Cimarron, Chevrolet Celebrity, Corsica and Beretta
Chevrolet	Camaro	V-6	2838	137@4900	217 Nm@3900	Also available in the Pontiac Firebird
	Camaro	V-8	5001	172@4000	346 Nm@2400	Also available in the Pontiac Firebird
	Camaro	V-8	5733	233@4400	448 Nm@3200	Also available in the Pontiac Firebird GT/A
Oldsmobile	Delta 88	V-6	3791	165@4400	275 Nm@2400	Also available in the Olds 98 and Toronado
Pontiac	Sunbird GT	4	1998	167@5600	237 Nm@4000	Turbocharged, also available on the Pontiac Grand Am and Buick Skyhawk
	6000 STE	V-6	3121	137@4800	244 Nm@3600	
	Firebird Trans-Am	V-8	5001	193@4000	400 Nm@2800	
	Grand Prix	V-6	2838	132@4500	217 Nm@3600	
	Bonneville	V-6	2838	152@4400	271 Nm@2000	
	Bonneville	V-6	2838	165@5200	285 Nm@2000	

Weber single-point injection

Make	Model	Number of cylinders	Displacement in cc	HP(DIN)@RPM	Torque@RPM	Remarks
Fiat	Regata	4	1585	100@6000	128 Nm@4000	

Weber Marelli IAW sequential injection

Make	Model	Number of cylinders	Displacement in cc	HP(DIN)@RPM	Torque@RPM	Remarks
AC	Ace	4	1994	204@6000	276 Nm@4500	Ford-based engine, Turbocharged
Aston Martin	V-8 Volante	V-8	5341	309@5000	434 Nm@4000	Also available in the Lagonda
Ferrari	F 40	V-8	2936	478@7000	577 Nm@4000	4 valves per cyl., twin turbo with intercooling
Fiat	Croma	4	1995	120@5250	167 Nm@3300	
Ford	Sierra Cosworth	4	1994	204@6000	276 Nm@4500	Turbocharged
	Sierra RS 500	4	1994	227@6000	280 Nm@4500	Turbocharged
Lancia	Delta Turbo	4	1585	140@5500	191 Nm@3500	Turbocharged
	Delta HF 2000	4	1995	185@5300	304 Nm@3500	Turbo-intercooled
	Prisma	4	1995	115@5400	163 Nm@3250	
Maserati	Biturbo Si	V-6	2491	188@5000	321 Nm@3000	Twin Turbo
	228 i	V-6	2790	250@6000	373 Nm@3500	Twin Turbo
	Karif	V-6	2790	285@6000	432 Nm@4000	Twin turbo, 3 valves per cyl.

Metric conversion table

To convert from	Unit	To	Unit	Multiply by:	To convert from	Unit	To	Unit	Multiply by:
Torque	Nm		lb. ft.	0.736	Displacement	cc		ci	0.061
Pressure	bar		psi	14.7	Weight	kg		lb	2.205
Temperature	degrees celcius		degrees fahrenheit	1.8, then add 32	Velocity	km/h		mph	0.621
					Linear measure	mm		in	0.0394
Fuel mileage	liter/100 km		mpg	Divide by 235					

Index